The Direct Measurement of Cardinal Utility:

A New Theory of Value & Demand

By M G Benthall

BA(Hons), MSc(Open), CEng, CITP,
MIMechE, MIET, MBCS

Copies of this book are also held by:
The British Library
Bodleian Libraries of the University of Oxford
Cambridge University Library
The National Library of Scotland
The Library of Trinity College, Dublin
The National Library of Wales

To order additional copies of this book, contact Xlibris:

Phone: (+44) 0800 056 3182
www.xlibrispublishing.co.uk
e-mail: Orders@Xlibrispublishing.co.uk

To contact the author:
e-mail: mike.benthall@sky.com

Print information available on the last page.
Revision date 21 November 2019

Dedication

To my father: the late Lesley Norman Benthall DSC. It was his knowledge of and interest in politics and economics, particularly Keynesian economics, that first fired my passion for the subjects.

Preface

The ideas behind this work first occurred to me while I was studying the Open University third level economics course D319 'Understanding Economic Behaviour: Households, Firms and Markets' in 2001.

I first studied with the OU in 1976, completing an ordinary BA degree in 1978. My intention was to go on to read for honours but in the following year I secured my first real management job. I took over a large department in considerable chaos and was working all hours trying to bring some order to the operation. I simply could not find the time to continue my studies. My books were put away but I vowed to return to my studies when I retired.

In 2000 I finally retired and true to my vow took up my studies again, re-enrolling with the OU in 2001. Because I obtained my ordinary degree so long ago, I had not studied the prerequisite course for D319. The material for the second level prerequisite course had however been published in the form of a book 'Economics and changing economies' (M. Mackintosh et al, 1996). So, I bought the book and read relevant chapters along with my course material for D319. One of the early sections of the course concerned consumer theory. I therefore read chapter 3 of the book that dealt with these matters. In that chapter (Jane Wheelock, 1996) derived an equation for the consumption function from the data in Appendix I table 11, using linear regression.

Looking at table 11 it was immediately apparent to me that the data did not describe a linear function at all but a saturating function. My original qualifications were in engineering and as an engineer I came to economics armed with the engineer's tool kit of mathematics. I had seen this sort of relationship before, it arises widely in electronics and systems engineering. My prior knowledge allowed me to write down the equation describing the relationship by inspection, well more or less, after a little head scratching and consulting my engineering texts.

If the consumption function was a saturating function it meant that human wants were satiable and orthodox neoclassical consumer theory was wrong. I set about examining the consequences of this discovery and to my shock found the whole of the established theory of value and demand was undermined by this simple reinterpretation of the consumption function.

For two months in the middle of my D319 course I did not touch my course work (not a good idea), spending the whole time sketching out the rudiments of the present theory. By Christmas 2001 I had completed the first typed draft. A further year was spent working out the mathematics of chapters 1 to 8 and yet another year, using MATHCAD 2001 professional, analysing partial and general equilibrium. Though one could go on for years pursuing more obscure consequences of the revised consumption function (and perhaps I shall) the fundamental aspects of the work are now complete.

It is usual in a preface to acknowledge the help and support that others have given, contributing to the work. In truth however, except for a vast amount of error checking for which I am eternally grateful, I received very little until my book was finished. I wrote to the course team of D319 asking for their help in getting the work published. They were unable to help me and may have thought I was mad. Surely, no one could accomplish so many startling results in what is after all a relatively short work? Many may still think I am mad. To those that do, I commend chapter 11, which derives a general equilibrium solution for all markets and chapter 10 that proves the uniqueness of partial equilibrium solutions in all, not just competitive markets. These results alone show this is a deadly serious work.

Even my wife, normally my strongest support in academic endeavour, was quite hostile initially thinking if I was not careful, I would fail my exams. She however eventually accepted the work was important to me and devoted a great deal of her time finding spelling errors the spell checker did not find and improving the grammar and punctuation. As an arts graduate of York University, she writes very much better than me and would have further improved the grammar and doubtless improved the style of the writing too had I not stubbornly rejected many of her suggestions concerning style.   Consequently, she wishes to disassociate herself from any residual weaknesses in my English, particularly my occasional use of very long sentences.

The Sleaford branch of the Lincolnshire County Library Service located and obtained nearly all the sources of reference used in this work, some from as far afield as Switzerland and America. I would like to thank the staff of Sleaford Library for all their help and advice.

My brother in law John Ryder, a Cambridge first, checked most of the mathematics in chapters 1 to 8 and some of the later sections, suggesting where the derivation was unsound. He was also intrigued by my combination of analytical and numerical approaches. The mathematics in later chapters was mainly developed using MATHCAD, making errors in the mathematics less likely. His brother Paul Ryder a graduate of the London School of Economics checked the economics as far as was practicable given that much of it was revolutionary and heretical. He raised some doubts and suggested some corrections that I have addressed. He also praised my work and offered useful advice on how I might get it published and gain acceptance of my ideas. One of my OU mathematics tutors, Dr Gareth Harries, an Oxford first, suggested a line of enquiry, which might explain why the general equilibrium solution I had obtained actually worked. This led indirectly to the solution included here. The Open University's Professor of Economics Dr Paul Anand gave me helpful advice on publication and agreed to briefly review my final draft. My thanks are due to all of these for their help and advice. Many pitfalls were avoided and errors removed as a result. All that remain are entirely my responsibility.

Mike Benthall
Original Preface January, 2004.
Revised Preface August, 2018

Contents

Table of Figures

List of Equations

$$q_x = (\alpha_x y e^{-y/\lambda} + \beta_x)/ P_x$$

$$D_x = (\sum_{1}^{N} (\alpha_x y e^{-y/\lambda} + \beta_x))/ P_x$$

$$D_x = (a_x Y e^{-Y/N\lambda} + b_x)/ P_x$$

$$P_x = (a_x Y e^{-Y/N\lambda} + b_x)/ D_x$$

$$\frac{dP_x}{dD_x} = (D_x \frac{dV_x}{dD_x} - V_x)/ D_x^2$$

$$\frac{dP_x}{dD_x} = - \frac{P_x}{D_x}$$

$$\varepsilon_{Px} = -1$$

$$P_X = A_X \cdot \frac{V_X}{D_X} + B_X$$

$$C_D = \frac{C_F}{q_x} + C_V + C_E \cdot q_x$$

$$\frac{dZ_x}{dD_x} = P_x + D_x \frac{dP_x}{dD_x} - C_D$$

$$0 = P_x + \frac{V_x dP_x}{P_x dD_x} - C_D$$

$$P_x = \frac{C_D \pm \sqrt{(C_D^2 - 4 V_x \frac{dP_x}{dD_x})}}{2}$$

$$D_x = \frac{2V_x}{C_D \pm \sqrt{(C_D^2 - 4 V_x \frac{dP_x}{dD_x})}}$$

$$f_5(D_x) = (f_1(D_x) - f_2(D_x))D_x$$

$$V_{SX} = V_x \log_e D_x - V_x$$

$$R_L = q_B \cdot \frac{r \cdot (1 + r)^n}{\left[(1 + r)^n - 1 \right]}$$

$$q_B = \sqrt{\alpha_B \cdot y \cdot e^{\frac{-y}{\lambda}} \cdot \frac{\left[(1 + r)^n - 1 \right]}{12 r \cdot (1 + r)^n}}$$

$$R_{Lx} = \sqrt{\alpha_{Bx} \cdot y \cdot e^{\frac{-y}{\lambda}} \cdot \frac{r \cdot (1 + r)^n}{12\left[(1 + r)^n - 1\right]}}$$

$$D_{Bx} = \sqrt{a_{Bx} \cdot Y \cdot e^{\frac{-Y}{N \cdot \lambda}} \cdot \frac{\left[(1 + r)^n - 1\right]}{12r \cdot (1 + r)^n}}$$

$$P_x = C_D \log_e D_x + \frac{V_x}{D_x}$$

$$P_x = C_V \cdot \log_e D_x + \frac{(V_x - C_F)}{D_x} + C_E (D_x - 1) + C_F$$

$$P_x = t_c \cdot C_D \log_e D_x + \frac{V_x}{D_x}$$

$$D_x = \frac{2V_x}{C_D}$$

$$P_x = C_D \log_e D_x + \frac{(e^{2y_{AV}/\lambda})V_x}{D_x}$$

$$\Delta_{s-c} \frac{dU_x}{dy} = 2a_x \left(\sinh(y/|\lambda|) + \frac{y \cosh(y/|\lambda|)}{|\lambda|} \right)$$

$$\Delta_{s-c} \frac{dU_x}{dy} = a_x e^{y/|\lambda x|} + \frac{a_x y e^{y/|\lambda x|}}{\lambda_x} - a_x e^{-y/\lambda} + \frac{a_x y e^{-y/\lambda}}{\lambda}$$

$$y_x = \left[\lambda_x \cdot W \left[\frac{-1}{2 \cdot \alpha_x \cdot \lambda_x} \cdot \left(-C_D \cdot \exp\left(\frac{-3}{2}\right) + 2 \cdot \beta_x \right) \right]^{-\lambda_x} \right]$$

$$P_x = V_x - C_D \log_e D_x - \frac{V_x}{D_x}$$

$$\frac{dv}{dP_r} = \sum_{x=1}^{r-1} \left(P_x \frac{dq_x}{dP_r} + q_x \frac{dP_x}{dP_r} \right) + q_r + P_r \frac{dq_r}{dP_r} + \sum_{x=r+1}^{n} \left(P_x \frac{dq_x}{dP_r} + q_x \frac{dP_x}{dP_r} \right)$$

$$\frac{1}{P_A} \frac{dU_A}{dq_A} = \frac{1}{P_B} \frac{dU_B}{dq_B}$$

$$\frac{P_B}{P_A} = \frac{q_B dP_B dq_A}{q_A dP_A dq_B}$$

$$B_{AB} = P_A q_A + P_B q_B$$

$$\frac{dq_A}{dq_B} = \frac{-P_B}{P_A}$$

$$P_1 + D_1 \cdot \frac{d}{dD_1} P_1 + D_2 \cdot \frac{d}{dD_1} P_2 + P_2 \cdot \frac{d}{dD_1} D_2 = 0$$

$$D_1 \cdot \frac{d}{dD_1} P_1 = -D_2 \cdot \frac{d}{dD_1} P_2$$

$$\int P_2 \, dD_2 = -\int P_1 \, dD_1$$

$$V_2 \cdot \ln(D_2) = -V_1 \cdot \ln(D_1) + k$$

$$D_2 = \exp\left[\frac{\left(-V_1 \cdot \ln(D_1) + k\right)}{V_2}\right]$$

$$D_2 = \frac{B}{P_2} - \frac{P_1}{P_2} \cdot D_1$$

$$D_2 = k_1 \cdot \exp\left[\frac{\left(V_1 \ln(D_1)\right)}{V_2}\right] + k_2$$

$$\int \frac{V_2}{D_2} \, dD_2 = -\int C_1 \cdot \ln(D_1) + \frac{V_1}{D_1} \, dD_1$$

$$V_2 \cdot \ln(D_2) = -C_1 \cdot D_1 \cdot \ln(D_1) + C_1 \cdot D_1 - V_1 \cdot \ln(D_1) + k$$

$$D_2 = \exp\left[\frac{\left(-C_1 \cdot D_1 \cdot \ln(D_1) + C_1 \cdot D_1 - V_1 \cdot \ln(D_1) + k\right)}{V_2}\right]$$

$$D_2 = k_1 \cdot \exp\left[\frac{\left(-C_1 \cdot D_1 \cdot \ln(D_1) + C_1 \cdot D_1 - V_1 \cdot \ln(D_1)\right)}{V_2}\right] + k_2$$

$$D_1 = \frac{2V_1}{C_1}$$

$$D_1 := \text{root}\left[\exp\left[\frac{\left(-C_1 \cdot D_1 \cdot \ln(D_1) + C_1 \cdot D_1 - V_1 \cdot \ln(D_1) + k\right)}{V_2}\right] - 1, D_1\right]$$

$$\int \frac{V_2}{D_2}\, dD_2 = -\int V_1 - C_1 \cdot \ln(D_1) - \frac{V_1}{D_1}\, dD_1 \qquad \text{Equation } 61 \text{ Page } 119$$

$$V_2 \cdot \ln(D_2) = -V_1 \cdot D_1 + C_1 \cdot D_1 \cdot \ln(D_1) - C_1 \cdot D_1 + V_1 \cdot \ln(D_1) + k \qquad \text{Equation } 62 \text{ Page } 119$$

$$D_2 = \exp\left[\frac{\left(-V_1 \cdot D_1 + C_1 \cdot D_1 \cdot \ln(D_1) - C_1 \cdot D_1 + V_1 \cdot \ln(D_1) + k\right)}{V_2}\right] \qquad \text{Equation } 63 \text{ Page } 119$$

$$D_2 = k_1 \cdot \exp\left[\frac{\left(-V_1 \cdot D_1 + C_1 \cdot D_1 \cdot \ln(D_1) - C_1 \cdot D_1 + V_1 \cdot \ln(D_1)\right)}{V_2}\right] + k_2 \qquad \text{Equation } 64 \text{ Page } 130$$

$$\int C_2 \cdot \ln(D_2) + \frac{V_2}{D_2}\, dD_2 = -\int C_1 \cdot \ln(D_1) + \frac{V_1}{D_1}\, dD_1 \qquad \text{Equation } 65 \text{ Page } 130$$

$$C_2 \cdot D_2 \cdot \ln(D_2) - C_2 \cdot D_2 + V_2 \cdot \ln(D_2) = -C_1 \cdot D_1 \cdot \ln(D_1) + C_1 \cdot D_1 - V_1 \cdot \ln(D_1) + k$$

$$\text{Equation } 66 \text{ Page } 130$$

$$D_1 = 1, 1.001..\frac{2V_1}{C_1} \qquad\qquad TOL = 10^{-12} \qquad\qquad D_2 = \frac{V_2}{C_2}$$

$$f_1(D_1) = \text{root}\left[C_2 \cdot D_2 \cdot \ln(D_2) - C_2 \cdot D_2 + V_2 \cdot \ln(D_2) - \left(-C_1 \cdot D_1 \cdot \ln(D_1) + C_1 \cdot D_1 - V_1 \cdot \ln(D_1) + k\right), D_2\right]$$

$$\text{Equations } 67 \quad \text{Page } 138$$

$$D_2 = \frac{k_1}{\frac{C_2}{V_2} \cdot \ln\left(\frac{V_2}{C_2}\right)} \cdot W\left[\frac{C_2}{V_2} \cdot \ln\left(\frac{V_2}{C_2}\right) \cdot \exp\left[\frac{\left(V_2 - C_1 \cdot D_1 \cdot \ln(D_1) + C_1 \cdot D_1 - V_1 \cdot \ln(D_1)\right)}{V_2}\right]\right] + k_2$$

$$\text{Equation } 68 \quad \text{Page..138}$$

$$f_3(D_1) := \frac{k_3}{\frac{V_2}{C_2} \cdot \ln\left(\frac{E \cdot V_2}{C_2}\right)} \cdot W\left[\frac{C_2}{V_2} \cdot \ln\left(\frac{E \cdot V_2}{C_2}\right) \cdot \exp\left[\frac{\left(E \cdot V_2 - C_1 \cdot D_1 \cdot \ln(D_1) + C_1 \cdot D_1 - V_1 \cdot \ln(D_1)\right)}{V_2}\right]\right] + k_4$$

$$\text{Equation } 69 \quad \text{Page } 140$$

$$\int C_2 \cdot \ln(D_2) + \frac{V_2}{D_2}\, dD_2 = -\int V_1 - C_1 \cdot \ln(D_1) - \frac{V_1}{D_1}\, dD_1 \qquad \text{Equation } 70 \text{ Page } 147$$

$$C_2 \cdot D_2 \cdot \ln(D_2) - C_2 \cdot D_2 + V_2 \cdot \ln(D_2) = -V_1 \cdot D_1 + C_1 \cdot D_1 \cdot \ln(D_1) - C_1 \cdot D_1 + V_1 \cdot \ln(D_1) + k$$

Equation   71   Page   147

$$D_1 = 1, 1.001.. \frac{2V_1}{C_1} \qquad TOL = 10^{-11} \qquad D_2 = \frac{V_2}{C_2}$$

$$f_1(D_1) = root\left[C_2 \cdot D_2 \cdot \ln(D_2) - C_2 \cdot D_2 + V_2 \cdot \ln(D_2) - \left(-V_1 \cdot D_1 + C_1 \cdot D_1 \cdot \ln(D_1) - C_1 \cdot D_1 + V_1 \cdot \ln(D_1) + k\right), D_2 \right]$$

Equations   72   Page   147

$$C_2 \cdot D_2 \cdot \left[\ln\left(\frac{V_2}{C_2}\right) + \frac{\left(D_2 - \frac{V_2}{C_2}\right)}{D_2} \right] - C_2 \cdot D_2 + V_2 \cdot \ln(D_2) = -V_1 \cdot D_1 + C_1 \cdot D_1 \cdot \ln(D_1) - C_1 \cdot D_1 + V_1 \cdot \ln(D_1)$$

Equation   73   Page   152

$$D_2 = \frac{k_1}{\frac{C_2}{V_2} \cdot \ln\left(\frac{V_2}{C_2}\right)} \cdot W\left[\frac{C_2}{V_2} \cdot \ln\left(\frac{V_2}{C_2}\right) \cdot \exp\left[\frac{\left(V_2 - V_1 \cdot D_1 + C_1 \cdot D_1 \cdot \ln(D_1) - C_1 \cdot D_1 + V_1 \cdot \ln(D_1)\right)}{V_2} \right] \right] + k_2$$

Equation   74   Page   152

$$f_3(D_1) = \frac{k_3}{\frac{C_2}{V_2} \cdot \ln\left(E \cdot \frac{V_2}{C_2}\right)} \cdot W\left[\frac{C_2}{V_2} \cdot \ln\left(E \cdot \frac{V_2}{C_2}\right) \cdot \exp\left[\frac{\left(E \cdot V_2 - V_1 \cdot D_1 + C_1 \cdot D_1 \cdot \ln(D_1) - C_1 \cdot D_1 + V_1 \cdot \ln(D_1)\right)}{V_2} \right] \right] + k_4$$

Equation   75   Page   153

$$\int V_2 - C_2 \cdot \ln(D_2) - \frac{V_2}{D_2} \, dD_2 = -\int V_1 - C_1 \cdot \ln(D_1) - \frac{V_1}{D_1} \, dD_1$$

Equation   76   Page   161

$$V_2 \cdot D_2 - C_2 \cdot D_2 \cdot \ln(D_2) + C_2 \cdot D_2 - V_2 \cdot \ln(D_2) = -V_1 \cdot D_1 + C_1 \cdot D_1 \cdot \ln(D_1) - C_1 \cdot D_1 + V_1 \cdot \ln(D_1) + k$$

Equation   77   Page   162

$$D_1 = 1, 1.001.. \frac{2V_1}{C_1} \qquad TOL = 10^{-12} \qquad D_2 = \frac{V_2}{C_2}$$

$$f_1(D_1) := root\left[V_2 \cdot D_2 - C_2 \cdot D_2 \cdot \ln(D_2) + C_2 \cdot D_2 - V_2 \cdot \ln(D_2) - \left(-V_1 \cdot D_1 + C_1 \cdot D_1 \cdot \ln(D_1) - C_1 \cdot D_1 + V_1 \cdot \ln(D_1) + k\right), D_2 \right]$$

Equation   78   Page   162

$$V_2 \cdot D_2 - C_2 \cdot D_2 \cdot \left[\ln\left(\frac{V_2}{C_2}\right) + \frac{\left(D_2 - \frac{V_2}{C_2}\right)}{D_2} \right] + C_2 \cdot D_2 - V_2 \cdot \ln(D_2) = -V_1 \cdot D_1 + C_1 \cdot D_1 \cdot \ln(D_1) - C_1 \cdot D_1 + V_1 \cdot \ln(D_1)$$

Equation 79 Page   165

$$D_2 = \frac{k_1 \cdot V_2}{\left(-V_2 + C_2 \cdot \ln\left(\frac{V_2}{C_2}\right)\right)} \cdot W\left[\frac{\left(-V_2 + C_2 \cdot \ln\left(\frac{V_2}{C_2}\right)\right)}{V_2} \cdot \exp\left[\frac{\left(V_2 + V_1 \cdot D_1 - C_1 \cdot D_1 \cdot \ln(D_1) + C_1 \cdot D_1 - V_1 \cdot \ln(D_1)\right)}{V_2}\right]\right] + k_2$$

$$\frac{V_2^2}{C_2} - V_2 \cdot \ln(D_2) + V_2 - V_2 \cdot \ln(D_2) = -V_1 \cdot D_1 + C_1 \cdot D_1 \cdot \ln(D_1) - C_1 \cdot D_1 + V_1 \cdot \ln(D_1)$$

$$D_2 = k_3 \cdot \exp\left[\frac{\left(\frac{V_2^2}{C_2} + V_2 + V_1 \cdot D_1 - C_1 \cdot D_1 \cdot \ln(D_1) + C_1 \cdot D_1 - V_1 \cdot \ln(D_1)\right)}{2 \cdot V_2}\right] + k_4$$

$$D_2 = \frac{k_1 \cdot V_2}{\left(V_2 - C_2 \cdot \ln\left(\frac{V_2}{C_2}\right)\right)} \cdot W\left[\frac{\left(V_2 - C_2 \cdot \ln\left(\frac{V_2}{C_2}\right)\right)}{V_2} \cdot \exp\left[\frac{\left(V_2 + V_1 \cdot D_1 - C_1 \cdot D_1 \cdot \ln(D_1) + C_1 \cdot D_1 - V_1 \cdot \ln(D_1)\right)}{V_2}\right]\right] + k_2$$

$$V = \sum_{x=1}^{n} P_x D_x$$

$$Y = \sum_{x=1}^{n} P_x D_x$$

$$\frac{dV}{dD_r} = \sum_{x=1}^{r-1}\left(P_x\frac{dD_x}{dD_r} + D_x\frac{dP_x}{dD_r}\right) + D_r\frac{dP_r}{dD_r} + P_r + \sum_{x=r+1}^{n}\left(P_x\frac{dD_x}{dD_r} + D_x\frac{dP_x}{dD_r}\right)$$

$$D_r\frac{dP_r}{dD_r} + \sum_{x=1}^{r-1} D_x\frac{dP_x}{dD_r} + \sum_{x=r+1}^{n} D_x\frac{dP_x}{dD_r} = 0$$

$$\frac{dV}{dD_r} = P_r + \sum_{x=1}^{r-1} P_x\frac{dD_x}{dD_r} + \sum_{x=r+1}^{n} P_x\frac{dD_x}{dD_r}$$

$$\int \frac{dV}{dD_r}, dD_r = \int P_r, dD_r + \sum_{x=1}^{r-1} \int P_x\frac{dD_x}{dD_r}, dD_r + \sum_{x=r+1}^{n} \int P_x\frac{dD_x}{dD_r}, dD_r$$

$$\int P_x\frac{dD_x}{dD_r}, dP_r = P_x D_x - \int D_x\frac{dP_x}{dD_r}, dD_r$$

$$\underline{V} = D.\underline{P}$$

Equation 91 Page 182

$$\underline{P} = D^{-1}.\underline{V}$$

Equation 92 Page 184

$$\underline{V} = -(n-2).\underline{P}$$

Equation 93 Page 191

$$D^{-1} = \frac{1}{2(n-2)} \cdot \begin{pmatrix} n-3 & \frac{-P_1}{P_2} & \frac{-P_1}{P_3} & \cdots & \frac{-P_1}{P_{n-1}} & \frac{-P_1}{P_n} \\[2mm] \frac{-P_2}{P_1} & n-3 & \frac{-P_2}{P_3} & \cdots & \frac{-P_2}{P_{n-1}} & \frac{-P_2}{P_n} \\[2mm] \frac{-P_3}{P_1} & \frac{-P_3}{P_2} & n-3 & \cdots & \frac{-P_3}{P_{n-1}} & \frac{-P_3}{P_n} \\[2mm] \cdot & \cdot & \cdot & \cdots & \cdot & \cdot \\ \cdot & \cdot & \cdot & & \cdot & \cdot \\ \cdot & \cdot & \cdot & & \cdot & \cdot \\ \cdot & \cdot & \cdot & & \cdot & \cdot \\ \cdot & \cdot & \cdot & \cdots & \cdot & \cdot \\[2mm] \frac{-P_n}{P_1} & \frac{-P_n}{P_2} & \frac{-P_n}{P_3} & \cdots & \frac{-P_n}{P_{n-1}} & n-3 \end{pmatrix}$$

Equations 94 Page 193

$$D^{-1} = \frac{1}{2(n-2)} \cdot \begin{pmatrix} n-3 & \frac{d}{dD_1}D_2 & \frac{d}{dD_1}D_3 & \cdots & \frac{d}{dD_1}D_{n-1} & \frac{d}{dD_1}D_n \\[2mm] \frac{d}{dD_2}D_1 & n-3 & \frac{d}{dD_2}D_3 & \cdots & \frac{d}{dD_2}D_{n-1} & \frac{d}{dD_2}D_n \\[2mm] \frac{d}{dD_3}D_1 & \frac{d}{dD_3}D_2 & n-3 & \cdots & \frac{d}{dD_3}D_{n-1} & \frac{d}{dD_3}D_n \\[2mm] \cdot & \cdot & \cdot & \cdots & \cdot & \cdot \\ \cdot & \cdot & \cdot & \cdots & \cdot & \cdot \\ \cdot & \cdot & \cdot & \cdots & \cdot & \cdot \\ \cdot & \cdot & \cdot & \cdots & \cdot & \cdot \\ \cdot & \cdot & \cdot & \cdots & \cdot & \cdot \\[2mm] \frac{d}{dD_n}D_1 & \frac{d}{dD_n}D_2 & \frac{d}{dD_n}D_3 & \cdots & \frac{d}{dD_n}D_{n-1} & n-3 \end{pmatrix}$$

$$\lim_{n \to \infty} \frac{1}{2(n-2)} \cdot \frac{d}{dD_r} D_x \to 0$$

if n is large $\dfrac{1}{2(n-2)} \cdot \dfrac{d}{dD_r} D_x \approx 0$

Equations 95 Page 194

$$\underline{P} = \frac{1}{2(n-2)} \cdot \begin{pmatrix} n-3 & \frac{-P_1}{P_2} & \frac{-P_1}{P_3} & \cdots \cdots & \frac{-P_1}{P_{n-1}} & \frac{-P_1}{P_n} \\ \frac{-P_2}{P_1} & n-3 & \frac{-P_2}{P_3} & \cdots \cdots & \frac{-P_2}{P_{n-1}} & \frac{-P_2}{P_n} \\ \frac{-P_3}{P_1} & \frac{-P_3}{P_2} & n-3 & \cdots \cdots & \frac{-P_3}{P_{n-1}} & \frac{-P_3}{P_n} \\ \cdot & \cdot & \cdot & \cdots \cdots & \cdot & \cdot \\ \cdot & \cdot & \cdot & & \cdot & \cdot \\ \cdot & \cdot & \cdot & & \cdot & \cdot \\ \cdot & \cdot & \cdot & & \cdot & \cdot \\ \cdot & \cdot & \cdot & & \cdot & \cdot \\ \frac{-P_n}{P_1} & \frac{-P_n}{P_2} & \frac{-P_n}{P_3} & \cdots \cdots & \frac{-P_n}{P_{n-1}} & n-3 \end{pmatrix} \cdot -(n-2) \cdot \begin{pmatrix} P_1 \\ P_2 \\ P_3 \\ \cdot \\ \cdot \\ \cdot \\ \cdot \\ P_n \end{pmatrix}$$

Equation 96 Page 196

$$D_S = \frac{1}{C_D} \cdot \frac{(V_S - V_C)}{W\left[\frac{1}{C_D} \cdot (V_S - V_C)\right]}$$

Equation 97 Page 198

$$D_G = \exp\left[\left[W\left[\frac{-(V_G + V_C)}{C_D} \cdot \exp\left(\frac{-V_C}{C_D}\right)\right] + \frac{V_C}{C_D}\right]\right]$$

Equation 98 Page 199

$$V_{ST} = \sum_{x=1}^{m} (V_x \log_e D_x - V_x)$$

Equation 99 Page 210

$$Y = \mu \cdot N$$

Equation 100 Page 213

$$U = \mu \cdot N \cdot A \cdot e^{\frac{-\mu}{h \cdot \lambda}}$$

Equation 101 Page 214

$$Y_A = k \cdot \sum_{i=0}^{b-1} n_i \cdot y_i$$

$$U_A = k \cdot \sum_{i=0}^{b-1} n_i \cdot y_i \cdot A \cdot e^{\frac{-y_i}{h \cdot \lambda}}$$

$$\delta U = \mu \cdot N \cdot A \cdot e^{\frac{-\mu}{h \cdot \lambda}} - k \cdot \sum_{i=0}^{b-1} n_i \cdot y_i \cdot A \cdot e^{\frac{-y_i}{h \cdot \lambda}}$$

$$n_t = \frac{N}{\sigma \cdot \sqrt{2\pi}} \cdot \exp\left[\frac{-1}{2 \cdot \sigma^2} \left(y_t - \mu \right)^2 \right]$$

$$Y_t = \int_{-\infty}^{\infty} \frac{N}{\sigma \cdot \sqrt{2 \cdot \pi}} \cdot \exp\left[\frac{-1}{2 \cdot \sigma^2} \cdot \left(y_t - \mu \right)^2 \right] \cdot y_t \, dy_t$$

or

$$Y_t = \mu \cdot \int_{-\infty}^{\infty} \frac{N}{\sigma \cdot \sqrt{2\pi}} \exp\left[\frac{-1}{2 \cdot \sigma^2} \cdot \left(y_t - \mu \right)^2 \right] dy_t$$

$$U_t = \int_{-\infty}^{\infty} \frac{N}{\sigma \cdot \sqrt{2\pi}} \exp\left[\frac{-1}{2 \cdot \sigma^2} \cdot \left(y_t - \mu \right)^2 \right] \cdot y_t \cdot A \cdot e^{\frac{-y_t}{h \cdot \lambda}} \, dy_t$$

$$\delta U = \int_{-h \cdot \lambda}^{h \cdot \lambda} \frac{N}{\sigma \cdot \sqrt{2\pi}} \cdot \exp\left[\frac{-1}{2 \cdot (\sigma)^2} \left(y_t - \mu \right)^2 \right] \cdot y_t \cdot A \cdot e^{\frac{-y_t}{h \cdot \lambda}} \, dy_t - k \cdot \sum_{i=0}^{b-1} n_i \cdot y_i \cdot A \cdot e^{\frac{-y_i}{h \cdot \lambda}}$$

$$c_t = \int_{-h \cdot \lambda}^{y_t} \frac{N}{\sigma \cdot \sqrt{2 \cdot \pi}} \cdot \exp\left[\frac{-1}{2 \cdot \sigma^2} \cdot \left(y_t - \mu \right)^2 \right] dy_t$$

$$c_i = k \cdot \sum_{i=0}^{b-1} n_i$$

$$y_t = \text{qnorm}(p, \mu, \sigma)$$

$$p = \frac{c_t}{N}$$

Equations   111   Page   226

List of Tables

Errata

To be used to collect all errors reported to the publisher.

Chapter 1 Introduction

For over a century, certainly since the work of (Vilfredo Pareto, 1909), economists have held that utility could not be measured or compared. Consequently, a whole system of analysis has developed based on the principle of ordinal utility that forms the present orthodoxy of the discipline.

This approach, as we shall see, is not entirely satisfactory and is merely the latest attempt to deal with an ancient conundrum of economics, the problem of the theory of value.   Since the foundation of the 'modern' discipline (Adam Smith, 1776), economists have grappled with three different notions of value: utility, price and cost. While some understanding of the relationship between these three concepts has emerged, the issue has never been entirely resolved.

This essay seeks to remove the principal stumbling block that has prevented the development of a 'complete' theory of value by showing how cardinal utility can be measured and marginal utility calculated. A whole new approach to the subject then becomes possible and a new theory of demand can be derived.

This theory is principally a neoclassical theory though we have become so enmeshed in the spiders' webs of indifference curves that much of it will seem quite unfamiliar. I nevertheless describe the theory as neoclassical since it is an analytical, reductionist, optimising theory based on greatly simplified formal models similar to the current orthodox neoclassical system.

Many of the assumptions underlying current neoclassical theory are however relaxed or shown to be invalid. Nearly all the results of current theory remain consistent with the new theory though a few are shown to be incorrect and many startling new results can be derived. The new theory also enables some points of dispute between neoclassical and institutionalist economics to be reconciled.

The ability to measure and calculate utility enables earlier utility theories, particularly those of (Alfred Marshal, 1890), to be at least partially rehabilitated. The same can be said of utilitarian theories, notably those of (Jeremy Bentham, 1789).   Though the new theory of value has much in common with these earlier theories it will nevertheless be shown to be distinct.

The theories of Vilfredo Pareto are not entirely abandoned but become of less importance.   Issues of redistribution, so long 'marginalized' by Pareto's concepts of optimality and efficiency, can once again be approached rigorously when we are armed with means of making quantifiable interpersonal comparisons of utility based on behavioural parameters. This enables new conditions for maximising social welfare to be established.

Achieving a greater precision in economic calculation is one of the objectives of the development of the new theory. As a consequence, this essay contains a great deal of mathematics. Economics, particularly undergraduate economics, often studiously avoids all 'advanced' mathematics particularly calculus. This is an error. It is however my intention to make the new theory as accessible as possible, though some of the mathematics developed here is entirely new. Nevertheless, the mathematical tools employed have been kept as simple as possible consistent with achieving the desired results. Leibniz notation is used in preference to functional notation, except where the latter is unavoidable, simply because the former is more widely understood. Similarly, where simplicity requires, mathematical rigour has been sacrificed to promote greater accessibility. All the results obtained here are however capable of rigorous proof.

Originally the new theory was not considered to be provable, though some particular results could be held to have been proven. As the work progressed however more and more of the conjectures of the theory have been proven to be correct. In its present form the theory rests substantially on proven results and on very little conjecture. This will not however become clear until towards the end of the essay.

The work is almost entirely theoretical. Hardly any empirical evidence is presented, though the starting point for developing the new theory is derived from empirical evidence. Similarly, it contains very little scholarship except to acknowledge the origins of some important ideas. The methodology is deductive, supported by mathematical derivation and illustrated with examples.

Much of the theory could be simply tested, though almost all the important relationships are non-linear and therefore unsuitable for normal econometric approaches based on linear regression. To econometricians seeking to test the theory I commend the methods of non-linear regression or those used in the physical sciences involving linear regression of calculated values of dependent variables predicted by the theory against measured values of the same variables under observably identical conditions.

Chapter 2. The Consumption Function

The consumption function is often evaluated using linear regression resulting in an equation of the form:

$$v = \alpha y + \beta \hspace{4cm} \text{Equation 1}$$

Where v is the value of an individual's consumption
y is the income of that individual
α is a constant of proportionality
and   β is the constant intercept on the vertical axis.

Though more elaborate forms of the consumption function are sometimes used they are generally developed from the linear model and, as we shall see, even this basic relationship is actually a very good model of consumption.

The data given in Appendix I gives aggregate consumption and income for the UK over the period 1950 to 1992. Clearly the resulting consumption function is not a linear function. This is most apparent if one examines the savings ratio, which, though erratic, increases with income. A better fit to this data is given by a function of the form shown in figure 1. How this function was derived is explained below.

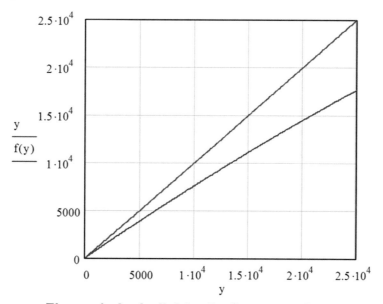

Figure 1. An Individual's Consumption Function

The red line of figure 1 represents income and the blue line the consumption function of an individual at increasing levels of income. The graph in figure 1 and most others in this work were originally drawn using the advanced mathematics computer application MATHCAD 2001 Professional. In this system the dependent variable is always represented on a graph by a function of the independent variable except where the relationship is defined wholly numerically by a matrix of co-ordinates. To keep a relatively uncluttered layout monetary units have been omitted throughout. Later revisions of this work were calculated using a later version of the software MATHCAD 15 released in 2010.

The form of the consumption function in figure 1 is a saturating growth function. In nature the most common form of growth, decay and saturating functions are all exponential functions. Indeed, the exponential function was developed in the first place with these uses in mind. Modelling the consumption function as a saturating exponential function will give rise to an equation of the following form:

$$v = \alpha y e^{-y/\lambda} + \beta \qquad\qquad \text{Equation 2}$$

Where v is the value of an individual's consumption
y is the income of that individual
α is a constant of proportionality
λ is a constant
and β is the constant intercept on the vertical axis.

We should take some trouble to define this function precisely since this will prove important later. The domain of y is positive real currency and zero which for the UK can be written symbolically as ($y \varepsilon £R_0^+$) where ε denotes 'is an element of'. The domain of α is the set of all real numbers and zero or ($\alpha \varepsilon R_0$). α is therefore dimensionless. The domain of β is the set of all real currency (positive and negative) and zero or for the UK ($\beta \varepsilon £R_0$). The domain of λ is the set of all real currency and zero or for the UK ($\lambda \varepsilon £R_0$). y/λ and $e^{-y/\lambda}$ are therefore dimensionless. Finally, the co-domain of v is the set of all positive real currency and zero or for the UK ($v \varepsilon £R_0^+$).

The form of the consumption function defined in equation 2 will be shown later, by the techniques of non-linear regression, to be a better model of consumption than the linear relationship of equation 1 generally used.

Disaggregating an Individual's Consumption Function

An individual's total consumption is composed of separate 'consumption functions' for each of the goods which that individual consumes. The disaggregation of an individual's consumption function can therefore be represented graphically as shown in figure 2.

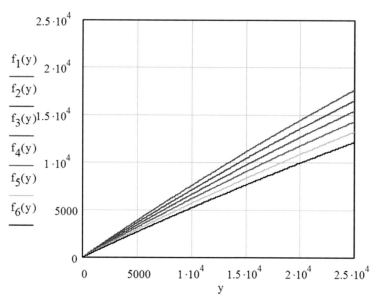

Figure 2. Disaggregation of an Individual's Consumption Function

The red line represents consumption of all the n goods in an individual's consumption set, the blue line consumption of n − 1 goods, the brown line consumption of n − 2 goods and so on. The consumption function for each of the goods 1 to n that an individual consumes is then given by:

$$v_x = \alpha_x y e^{-y/\lambda} + \beta_x$$ Equation 3

Where x is an arbitrary good.

Though extrapolating the full form of the aggregate consumption function must be considered to be at best uncertain because we have only ever been able to observe a small portion of the curve, this is certainly not the case for all individuals where a wide range of incomes can be 'earned' in almost all economies. We could therefore test that the full form of the consumption function of an individual, or for any particular good consumed by an individual, is given by a graph of the form shown below. Figure 3 shows the extended consumption function of an individual using the same function and parameters as figure 1 and the value of λ calculated later in the essay.

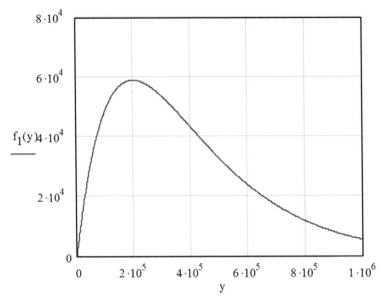

Figure 3. Full Form of an Individual's Consumption of Good x

In reality for goods that do not constitute a very great part of total consumption the trajectory of the curve is very low and flat, spanning a wide range of income.   Goods such as housing everywhere or staple foods in the third world will have higher values of α_x and probably β_x, hence the trajectory of the consumption function for these goods will be higher and less flat.

If the value of β_x is greater than zero, consumption of x will never fall below β_x. β_x thus represents priority. Goods for which β_x are greater than zero therefore constitute an individual's Pareto optimal minimum or permanent consumption set, in the consumer's own estimation.   The range of incomes over which such goods are consumed if 'preferences' do not change will therefore be infinite.

The implication of the form of the consumption function is however that consumption of all goods will eventually begin to fall if income rises to a sufficiently high level and the supply is sufficient. All goods will therefore eventually become inferior goods (in the conventional sense) at a sufficiently high level of income.

We can derive an individual's marginal propensity to consume good x by differentiating v_x with respect to y. Thus:

$$\frac{dv_x}{dy} = \alpha_x e^{-y/\lambda} - \frac{\alpha_x y e^{-y/\lambda}}{\lambda}$$

or $\qquad\qquad \frac{dv_x}{dy} = \alpha_x e^{-y/\lambda}(1 - y/\lambda)$ $\qquad\qquad$ Equation 4

From equations 2 and 4 an individual's aggregate marginal propensity to consume all goods is given by:

$$\frac{dv}{dy} = \alpha e^{-y/\lambda}(1 - y/\lambda) \qquad\qquad \text{Equation 5}$$

The form of an individual's marginal propensity to consume function is as shown in figure 4.

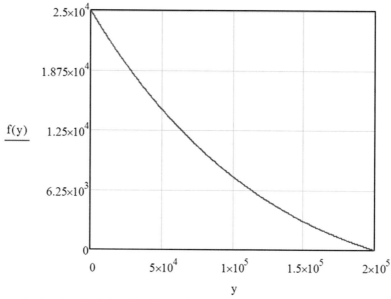

Figure 4. An Individual's Marginal Propensity to Consume Function

Though this graph is the result of differentiation with respect to income rather than demand the form of the graph substantially rehabilitates the 19th century view of marginal utility particularly that of (Alfred Marshall, 1890), suggesting we should review our sceptical opinion of 'The Law of Diminishing Marginal Utility'.

The 'Aggregate' Consumption Function for a Good

The aggregate consumption function for good **x** for the whole economy can be derived from the consumption functions of individual consumers by vector addition in two-dimensional vector space.   When we add consumption functions across the economy, we are effectively adding both consumption and income as shown schematically in Figure 5 below. Note this schematic graph unlike all other graphs in this work is not drawn using MATHCAD.

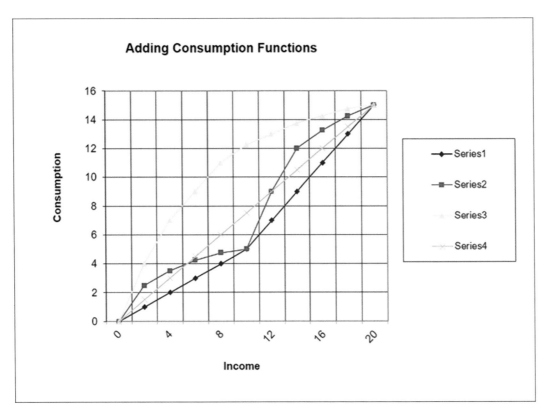

Figure 5. Adding Consumption In Two-Dimensional Vector Space.

The total market consumption of good x is thus given by the vector sum:

$$\begin{pmatrix} V_x \\ Y \end{pmatrix} = \begin{bmatrix} \sum_{i=1}^{N} \left(\alpha_x \cdot y_i \cdot \exp\left(\frac{-y_i}{\lambda} \right) + \beta_x \right) \\ \sum_{i=1}^{N} y_i \end{bmatrix}$$

Let

$$a_x \cdot Y \cdot \exp\left(\frac{-Y}{N\lambda} \right) = \sum_{i=1}^{N} \alpha_x \cdot y_i \cdot \exp\left(\frac{-y_i}{\lambda} \right)$$

then

$$a_x = \frac{\sum_{i=1}^{N} \alpha_x \cdot y_i \cdot \exp\left(\frac{-y_i}{\lambda} \right)}{\left(Y \cdot \exp\left(\frac{-Y}{N\lambda} \right) \right)}$$

and let:

$$b_x = \sum_{i=1}^{N} \beta_x$$

Where
Y is National Income
i is an individual consumer
N is the total population
y_i is the income of consumer i
a_x is the income or value weighted aggregate parameter α_x of all consumers for good x
b_x is the aggregate parameter β_x of all consumers for good x.
exp is the exponential function

The aggregate consumption function for good x is then given by:

$$V_x = a_x Y e^{-Y/N\lambda} + b_x \qquad \text{Equation 6}$$

and the aggregate marginal propensity to consume good x is therefore:

$$\frac{dV_x}{dY} = a_x e^{-Y/N\lambda}\left(1 - \frac{Y}{N\lambda}\right) \qquad \text{Equation 7}$$

These are results of the same form as equations 3 and 4 aggregated for the whole economy. Readers may be wondering why the population N now appears in the term $-Y/N\lambda$. This is because the income of N consumers is on the average N times the income of one consumer. Another way of looking at this is that the overall aggregate consumption function of a good contains a term representing average income. Since average income is given by:

$$y_{AV} = \frac{\sum y}{N}$$

and since $Y = \sum y$ then average income is given by Y/N.

The Aggregate Consumption Function

Finally the overall aggregate consumption function for all consumers of all goods can be calculated by adding the aggregate consumption functions of each good. This can be done algebraically since National Income is common to the aggregate consumption functions of all goods.

Therefore:
$$V = \sum_{1}^{n}(a_x Y e^{-Y/N\lambda} + b_x)$$

Where
and
V is overall aggregate consumption
n is the number of goods in the economy

This can be simplified to:
$$V = AYe^{-Y/N\lambda} + B \qquad \text{Equation 8}$$

Where
$$A = \sum_{1}^{n} a_x$$

and
$$B = \sum_{1}^{n} b_x$$

From equations 7 and 8 the marginal propensity to consume for the whole economy is then given by:

$$\frac{dV}{dY} = Ae^{-Y/N\lambda}\left(1 - \frac{Y}{N\lambda}\right) \qquad \text{Equation 9}$$

This is the marginal propensity to consume of Keynesian macro-economics. (John Maynard Keynes, 1936) called this function the aggregate demand function. In fact, as Keynes acknowledges, it is not a demand function at all but the marginal propensity to consume function. Many of the so-called parameters of economics can be evaluated as functions in this system yielding more general results. We can of course reformulate all the expressions in this system into parameterised form to facilitate local estimates of variables around particular points of interest on any of the functions. This approach may have many econometric uses.

Equation 9 gives a far more accurate way of evaluating the marginal propensity to consume than the decidedly crude method commonly used based on year on year changes. If we know National Income, total consumption and can estimate the value of λ we can calculate the value of 'A' and 'B' very precisely. A far more stable value for the marginal propensity to consume could then be calculated than is commonly used and therefore better estimates of the Keynesian multiplier could be made.

The value of the aggregate marginal propensity to consume will of course change with changing economic circumstances but these will be reflected in changing levels of income and consumption. A more accurate estimate could therefore be produced at any point in time using the latest National Income Accounts. Note that the marginal propensity to consume for the whole economy is derived from individual consumers' preferences for individual goods. This issue will be investigated further in the next chapter.

Chapter 3. A New Theory of Consumer Choice

So far we have merely derived a form of consumption function that is potentially a little more accurate than the function generally used, and shown how this function can be disaggregated and aggregated for each good and thence the whole economy. If that were the extent of the implications of this insight the new consumption function would be little more than a curiosity.

It transpires however that a whole new theory of consumer choice, value and demand can be derived from the new form of consumption function. The cumulative effect of this new theory is substantially to overturn (or at least limit the applicability of) the dominant concept of modern neoclassical microeconomics, the principle of ordinal utility. We will begin by examining the implications of the form of the new consumption function and the interpretation of its parameters. We will thence develop a new and simpler theory of consumer choice.

Properties of the Parameters α_x, β_x, α, β, a_x, b_x, A and B

The implication of a consumption function with a saturating form is that human wants are satiable both individually and in total. Such a consumption function could not possibly occur if human wants were insatiable. The first casualty of the new theory of value is therefore the neoclassical axiom of insatiability

The parameter α_x is a measure of an individual's preference for good x since it directly measures the initial proportion of income allocated to consumption of good x before satiation, represented by the term $e^{-y/\lambda}$, reduces the proportion of income devoted to its consumption. α_x must therefore be a behavioural parameter. When α_x is positive the good is a normal good. Inferior goods at incomes below λ must have a negative value for α_x since consumption falls with rising income.

The parameter β_x is a measure of an individual's priority given to good x. When β_x is positive x has high priority for the individual who will consume the good from borrowing even when income is zero. Such goods would include staple foods and any 'good' to which an individual is addicted. The value of β_x also directly measures borrowing made to finance consumption of good x.

When β_x is zero, consumption of a good will commence as soon as any income is earned. When β_x is negative, consumption will only commence at elevated levels of income. We can therefore define goods with a negative value of β_x as 'superior' or luxury goods. The more negative the value of β_x is for a good the higher the level of income must be before that good is consumed at all. β_x is therefore a measure of a good's relative superiority. Because some goods are superior, the number of goods in an individual's consumption set is therefore likely to rise as income rises. Note this definition does not imply that a superior good as defined here is the obverse of a conventionally defined inferior good.

Figure 6 below illustrates how, since consumption can 'never' be negative, a negative value of β_x 'delays' consumption until a minimum level of income is reached.

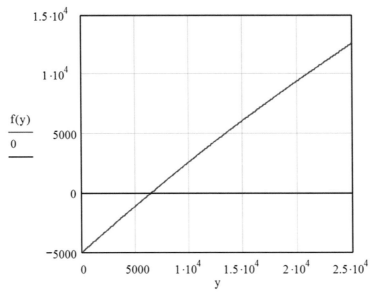

Figure 6. Effect of a Negative Value of β_x

The aggregate parameter α varies the initial proportion of income that an individual devotes to consumption in total.   It is therefore a behavioural parameter. When consumption is financed from pre-existing assets or initial endowment as well as income, the value of α will be greater than 1, otherwise the value of α will be less than or equal to 1. α must always have a value greater than zero if an individual's total consumption increases at all with income.

β is also a behavioural parameter. The aggregate parameter β must have a value greater than zero for a net borrower, less than zero for a net lender or equal to zero for a consumer who is neither a net lender nor a net borrower. If positive, it measures the total balance of borrowing of an individual over the period for which income is defined.   When income is zero β defines an individual's minimum level of consumption Pareto optimally in the consumer's own estimation. The value of β when income is zero therefore provides a Pareto optimal definition of relative poverty.

The values of α and β will differ from individual to individual depending upon the utility they derive from consumption overall and their attitude to their own personal financial security. Risk preferers will have higher values for both α and β than risk avoiders.

The parameter a_x therefore represents the income or value weighted aggregate preference of all consumers to consume good x, and the aggregate parameter A represents the income or value weighted aggregate preference of all consumers to consume all goods. Since aggregate consumption is almost invariably less than National Income in all economies, the parameter A directly represents the initial proportion of all income devoted to consumption.

The parameter b_x represents the aggregate priority of all consumers to consume good x, and aggregate consumer borrowing to finance good x. The parameter B represents aggregate consumer priority for all consumption of all goods and all borrowing to finance consumption.

Because of the circular flow of income, total net borrowing (borrowing − lending) for the whole world economy must be zero. For any National economy net borrowing could in principle be greater than or less than zero depending on the prevailing balance of trade. We have a reasonable estimate of the likely maximum value of B in the aggregate consumption function for the UK from the vertical intercept of a linear regression equation of the form of equation 1 applied to the data given in Appendix I by (Jane Wheelock, 1996). Dr Wheelock gave the vertical intercept of an aggregate consumption function of the form of equation 1 as £9,425.86. This figure is far too low to be due to a balance of payments deficit. The best interpretation is to view the figure obtained by Dr Wheelock as being due to noise. In any event the figure is likely to be lower when using a consumption function of the form of equation 8 because the effect of fitting a linear function to data drawn from a saturating function is to rotate the line of best fit clockwise about the centroid of the data points. This results in an apparent intercept that is higher than the true intercept and a reduced slope. These two arguments taken together make it reasonable to assume that the value of B in equation 8 is in principle equal to zero. We will however retain the form of equation 8 to include B for the time being until this presumption has been confirmed.

It should really be no surprise that the parameters of the consumption function directly represent preference and priority. Indeed, it would be more surprising if this were not the case. In the physical world the parameters of a functional relationship between variables generally have physical meaning. The real puzzle is how this obvious deduction concerning consumption has been overlooked for so long.

An individual's consumption function for a particular good thus provides a simple but exact specification of his preference for that good. This specification does not however satisfy the usual assumptions made in neoclassical ordinal utility theory. Equation 3 allows priority as well as preference to be given to all goods. Consumer preferences must therefore be lexicographical and as we shall see not necessarily consistent contrary to the assumptions of neoclassical consumer theory. That preferences may be lexicographical follows from assuming either a linear or a saturating exponential consumption function since the same parameters α_x and β_x could be considered to arise in either formulation. That preference is not consistent but varies with income follows from the saturating form of the aggregate consumption function. These conditions can thus be taken as universal and the assumptions of neoclassical theory must therefore be incorrect.

We will defer further development of the theory of consumer choice until we have derived expressions for an individual's demand function for reasons that will become apparent.

The Absolute Value of Money

The value of λ could also be a behavioural parameter but, since these equations must work for all currencies at all levels of inflationary prices, it seems more likely that the value of λ is related to the value of money. Since λ cannot perform this function and be a behavioural parameter we may assume it is not. Another reason for this assumption will be given later when we consider the value system of all consumers. Thus, we can write:

$$\frac{y_\$}{\lambda_\$} = \pm y_{ABS}$$

$$\frac{y_£}{\lambda_£} = \pm y_{ABS}$$

$$\frac{y_0}{\lambda_0} = \frac{y_t}{\lambda_t} = \pm y_{ABS}$$

$$\frac{|\lambda_\$|}{|\lambda_£|} = R_{\$/£}$$

and
$$\frac{\lambda_t}{\lambda_0} = \pm R_{t-0}$$
Equations 10

Where
- $y_\$$ is income expressed in \$s
- $y_£$ is income expressed in £s
- y_{ABS} is income expressed in absolute money
- $\lambda_\$$ is a constant divisor to convert \$s into absolute money
- $\lambda_£$ is a constant divisor to convert £s into absolute money
- λ_0 is a constant divisor to convert prices or incomes at time 0 into absolute money
- λ_t is a constant divisor to convert prices or incomes at time t into absolute money
- $R_{\$£}$ is the exchange rate of \$s per £s

and
- R_{t-0} is the inflation rate between time 0 and time t.

Similar relationships will exist for the € or any other currency.

λ will have quite a high absolute value as it determines the curvature, though not the slope, of the saturating consumption function since the permanent satiation of human wants is very gradual. Under certain circumstance that will be investigated later λ can have a negative value. This is why the inflation rate and the value of the £ and the \$ in equations 10 have a plus or minus sign. We will consider how to calculate the absolute value of λ in the next section of this essay.

Since λ is expressed in currency there is no money illusion implied by a consumption function in this form. v and y are both expressed in money terms and are on the average equally affected by inflation. y/λ is dimensionless which implies the units of absolute money are dimensionless. Unlike the purely linear consumption function the new model can properly represent both inflationary growth of prices and incomes and the real growth of income and consumption. The former has the effect of increasing the value of λ in proportion to any inflationary change in y or v whereas real growth leaves λ unchanged despite changes in y and v. This property will have implications for general equilibrium.

Money has three distinct functions: as a store of wealth, a medium of exchange, and as a measure of relative scarcity, utility or value. Each of these functions of money is reflected in the current theory and each can fail separately or together under particular circumstances that will be reviewed later in the essay.

Estimating the Value of λ

The value of λ is difficult to estimate because income and consumption data collected in National Income Accounts is subject to a great deal of random variation or noise. We know that λ must have a very high value because the satiation of human wants is very gradual, so gradual in fact that for generations the consumption function has been mistakenly assumed to be linear and generally modelled in that way.

To provide a reasonable estimate of the value of λ we therefore need to use data from a long time series to try and smooth out short-term fluctuations. The data in Appendix I gives official UK National Income and consumption statistics from 1950 to 1992 while Appendix II gives data on UK population. Taken together these provide reasonably reliable sources of data from which to estimate a value for λ. The UK has the longest continuous record of data of National Income, which helps in producing a stable estimate for λ.

Because the consumption function is a non-linear function that cannot be reduced to linear form we need to use the technique of non-linear regression. The mathematics of this technique are unfortunately not as simple as those used for linear regression. Furthermore, the functions provided in spreadsheets for non-linear regression are also unsuitable and unreliable in this instance. We will therefore make use of the advanced mathematics software, MATHCAD 2001 Professional, and a MATHCAD worksheet developed by the Open University 'Minimizing a function of n variables: least squares problems' (OU, 2002).

The number of data points available in the source data of Appendix I is unfortunately too great for MATHCAD to calculate the values of the parameters of equation 8 on a circa 2001 PC in a reasonable time. I have therefore selected the data for 1950 and every subsequent 5th year up to 1990 inclusive.

An 'equation' expressing residual errors was set up in the OU MATHCAD worksheet as shown below.

$$
r(x) := \begin{pmatrix}
1.25185x_1 \cdot \exp\left(\dfrac{-1.25185}{x_2}\right) + x_3 - 1.22649 \\[2ex]
1.42664x_1 \cdot \exp\left(\dfrac{-1.42664}{x_2}\right) + x_3 - 1.37136 \\[2ex]
1.69199x_1 \cdot \exp\left(\dfrac{-1.69199}{x_2}\right) + x_3 - 1.56735 \\[2ex]
1.96998x_1 \cdot \exp\left(\dfrac{-1.96998}{x_2}\right) + x_3 - 1.78493 \\[2ex]
2.17675x_1 \cdot \exp\left(\dfrac{-2.17675}{x_2}\right) + x_3 - 1.97873 \\[2ex]
2.53814x_1 \cdot \exp\left(\dfrac{-2.53814}{x_2}\right) + x_3 - 2.24580 \\[2ex]
2.85411x_1 \cdot \exp\left(\dfrac{-2.85411}{x_2}\right) + x_3 - 2.47185 \\[2ex]
3.09821x_1 \cdot \exp\left(\dfrac{-3.09821}{x_2}\right) + x_3 - 2.76742 \\[2ex]
3.80092x_1 \cdot \exp\left(\dfrac{-3.80092}{x_2}\right) + x_3 - 3.47527
\end{pmatrix}
$$

Where r(x) is a vector of residuals
x_1 is the parameter A in equation 8
x_2 is the parameter λ in equation 8
x_3 is the parameter B in equation 8

The numerical values 'represent' income and consumption in each of the years selected. Note the larger figure represents income in each case.

There are two ways we could define λ either with respect to the whole economy or with respect to an individual. I have chosen the latter. The source data however is defined for the whole economy. So as not to introduce any distortion into the data it is proposed to first calculate λ for the whole economy and thence to deduce λ for an individual.

The expected value of λ is of the order of at least 10^{11} times greater than the expected value of A, while the expected value of B is zero. This introduces a scaling problem into the model. To offset this, equation 8 has been divided by 10^{11} throughout. Income and consumption are therefore expressed in units of one hundred billion £s. This is the form in which the data appears in the vector r(x) above.

A grid search was carried out which revealed the value of B to be approximately zero. This confirms our expectation concerning net borrowing. To simplify the regression process the vector r(x) was therefore redefined as shown below, eliminating B which as you may recall was represented by x_3. The validity of this assumption will be confirmed later.

r(x) therefore becomes:

$$r(x) := \begin{pmatrix} 1.25185x_1 \cdot \exp\left(\dfrac{-1.25185}{x_2}\right) - 1.22649 \\[1.2em] 1.42664x_1 \cdot \exp\left(\dfrac{-1.42664}{x_2}\right) - 1.37136 \\[1.2em] 1.69199x_1 \cdot \exp\left(\dfrac{-1.69199}{x_2}\right) - 1.56735 \\[1.2em] 1.96998x_1 \cdot \exp\left(\dfrac{-1.96998}{x_2}\right) - 1.78493 \\[1.2em] 2.17675x_1 \cdot \exp\left(\dfrac{-2.17675}{x_2}\right) - 1.97873 \\[1.2em] 2.53814x_1 \cdot \exp\left(\dfrac{-2.53814}{x_2}\right) - 2.24580 \\[1.2em] 2.85411x_1 \cdot \exp\left(\dfrac{-2.85411}{x_2}\right) - 2.47185 \\[1.2em] 3.09821x_1 \cdot \exp\left(\dfrac{-3.09821}{x_2}\right) - 2.76742 \\[1.2em] 3.80092x_1 \cdot \exp\left(\dfrac{-3.80092}{x_2}\right) - 3.47527 \end{pmatrix}$$

The OU MATHCAD worksheet employs an iterative process that requires appropriate starting values. Suitable initial estimates for x_1 and x_2 were established using a 100x100 point grid search over the intervals (0.8, 1.0) and (1, 1000) respectively. The grid search returned approximate values for x_1 and x_2 of 0.932 and 90.91 respectively. Two methods of calculating the line of best fit of equation 8 to the data given above were employed: the Broyden-Fletcher-Goldfarb-Shanno BFGS method and the Gauss-Newton method. In both cases exact golden section line searches were used. Suitable stopping criteria were used to ensure six decimal place accuracy of the main iterative process and seven decimal place accuracy of line searches. Both methods reported the solution found to the required accuracy and both returned the values (0.93112, 91.99692).

This result gives a value for A of 0.931120 and for λ defined for the whole economy of £9.199692 x 10^{12} at 1990 prices. The UK Annual Abstract of Statistics 2001 gives the mid-point estimate of UK population in 1990 as 57,567,000. The value of λ for an individual is therefore estimated to be

£159,808 at 1990 prices. The compounded annualised Retail Price Index RPI in the UK for the period 1991 to 2000 inclusive derived from the data given in Appendix III is 34.9557%. The estimated value of λ for the UK at year 2000 prices is therefore £215,671.

This is a rather simplistic way of estimating λ for an individual since it applies to the separate consumption of every man, woman and child in the country irrespective of whether they are in receipt of a separate income. Depending on the purpose for which equation 8 is to be used it might be better to calculate the value of λ for each household, wage earner and benefit receiver or tax payer. We will however persist with our calculated value or an approximation to it throughout most of the remainder of this essay.

The resulting function produces a correlation coefficient of 0.997 and a value for R^2 of 0.993 that compares favourably with an R^2 value of only 0.948 obtained by (Jane Wheelock, 1996) from linear regression using all the original source data. Note however that though no distortion was intended (just an arbitrary selection of data points at equal time intervals) Dr Wheelock may also have obtained a better R^2 value for the chosen subset of data. This result confirms that a saturating form of the consumption function provides a better model than the customary linear relationship and that the value of B is zero. This may not seem to be much of an improvement but in fact it has reduced residual errors in R^2 arising from linear regression by (0.052 − 0.007)/0.052. That is by over 86.5%.

The precision with which these calculations were carried out should not be taken as indicating that we have determined λ to this precision. The main reason for using such accuracy is that the more precisely we attempt to calculate λ the more likely it becomes that the procedures we used would fail to converge to the required accuracy. The fact that both procedures did converge (using this data set at least) is in itself helpful in validating the theory. Treating the precision of the results with some caution therefore we could at least claim the value of λ for an individual (defined as we defined an individual) is therefore of the order of £216,000 or to simplify calculations approximately £200,000 at year 2000 prices. This seems an appropriate base line for the theory. Except where otherwise stated (for example where greater precision is necessary) the latter convenient figure has at times been used throughout the remainder of this essay making graphs simpler to interpret.

Savings

Savings are represented on the graph of figure 1 by the widening gap between the red income line and the blue consumption line. Economists generally treat savings as deferred consumption. The form of the saturating consumption function however suggests an alternative interpretation. Because the aggregate consumption function is a saturating function with a vertical intercept of zero, savings may be viewed as a residual part of income arising from satiating consumption. From this perspective the consumption function could be taken as directly measuring cardinal utility. The curvature of the consumption function is certainly of the correct form to represent utility. It is a very slowly saturating function displaying the expected property of diminishing marginal utility of income.

Ice cream is often given as an example of diminishing marginal utility. 'One ice cream is very nice but five may make you sick.' This is not the same phenomenon as that represented by the consumption function. In the ice cream example satiation is temporary. Because consumption is continuous or at least continual, the form of satiating consumption represented in figure 1 is permanent ongoing satiation, not just a temporary sufficiency.

Before adopting the precept that the consumption function directly measures utility however, we need to consider the orthodox explanation of why people save. One way of doing this is to consider consumers motive and opportunity to save. The poor may have a motive to save but they have little opportunity. The middle-class have both motive and opportunity but why do the rich save? They have great opportunity but what is their motive?

The middle-class's saving certainly fits the model of deferred consumption. Typically, their incomes peak between the ages of 45 and 55. Thereafter real income generally falls throughout the rest of life. Increased saving during the peak years therefore helps to smooth out the peaks and troughs of lifetime earnings. Middle-class families may no longer need to support their children by the time their earnings peak providing additional opportunities to save.

An individual's motivation for saving may be many and varied but substantial and prolonged saving seems to be aimed principally at improving long term financial security. The modest middle-class saver is likely to be risk averse and therefore reasonably safe forms of 'investment' will be preferred. Historically positive real interest rates in safe 'investments' are rare and positive interest net of tax and inflation even more so. For the middle-class therefore saving may involve a loss of value. Thus, saving does not always seem to be the behaviour we would expect from a neoclassical rational optimiser.

The wealthy, saving on a grander scale, generally need not suffer losses due to negative real interest rates because they can establish large diversified portfolios that spread their risk. On the average the performance of their 'investments' will tend to the growth rate of the economy as a whole. The rich however face higher marginal tax rates so even diversified 'investments', may involve a loss of value despite available tax breaks.

The rich, and particularly the very rich, do not have the motive of the middle class of improving financial security. Their future security is already assured but they continue to save even more than the middle-class. A rational optimiser might be expected to increase consumption once future security was guaranteed. This would certainly be the case if savings were also subject to diminishing marginal utility of income. In fact, the rich can be observed to do the opposite. That saving does not display diminishing marginal utility could be taken as a proof that savings are a residual.

The motives of the rich may include a desire to increase the inheritance of their heirs but even those without heirs continue to save ever-increasing amounts as income rises. One is compelled to the view that we save despite our rational interests and not because of them.

So why do we save? We save because our consumption is satiating as our income rises and no matter how we wriggle we cannot escape from this. Savings are just that part of income we cannot consume because our wants are satiating. The fact that we may gain other benefits from saving is just serendipity.

Value System

It follows from the above that in addition to the form given in equation 3 that gives consumption of a good as a function of income, the value of an individual's consumption of a good can be represented as the product of the quantity consumed and the price paid for the good.

Thus $$v_x = P_x q_x \qquad \text{Equation 11}$$

Where P_x is the market price
and q_x is the quantity of x an individual consumes

Irrespective of whether a firm supplying a good is a price taker, all consumers are almost invariably price takers (except occasionally in some auction markets) since there are generally far more buyers than sellers. P_x is therefore the market price whereas q_x is the individual's quantity consumed.

Since the saturating form of the consumption function for an individual good arises through satiating consumption or diminishing marginal utility of income then the consumption function directly represents the total utility arising from consumption of that good. We can therefore also write:

$$u_x = P_x q_x$$

or since $u_x = v_x$ $$u_x = \alpha_x y e^{-y/\lambda} + \beta_x \qquad \text{Equations 12}$$

Where u_x is the utility an individual consumer derives from consumption of good x

More generally we can define the utility derived from ownership or consumption of a good under the new theory of value as the opportunity cost of obtaining or retaining that good. The marginal utility an individual derives from the consumption of a good under the new theory of value can therefore be defined as the first derivative of the opportunity cost of obtaining or retaining that good with respect to the quantity consumed.

We can therefore simply derive the marginal utility an individual derives from the consumption of good x as:

$$\frac{du_x}{dq_x} = P_x + q_x\frac{dP_x}{dq_x} \qquad \text{Equations 13}$$

Note that this equation holds irrespective of the form of demand function that may apply to good x. For example, this equation holds even if good x is a speculative good.

Because consumers are almost invariably price takers then the price of a good is assumed not to change with the quantity that an individual consumes. The rate of change of price with respect to an individual's change of demand may therefore be assumed to be negligible, tending to zero and the marginal utility an individual derives from consumption of good x is thus given by:

$$\frac{du_x}{dq_x} \approx P_x \qquad \text{Equation 14}$$

Though all consumers will have different incomes and different preferences and priorities concerning the consumption of each of the goods in their consumption set, giving rise to a different quantity demanded q_x by each consumer, they all subscribe to the same system of values. Each consumer simply adopts the market price P_x, as equation 14 shows, that reflects the relative scarcity of a good, as the value they must ascribe to that good so that they correctly economise in its use. This is what we mean when we say all consumers are price takers. They are obliged to value goods as the market values them. The utility or value all consumers attach to all goods is therefore the same and collectively determined, whereas each consumer's preferences and priorities (and therefore their expenditure on each good) are subjectively determined.

Indeed, individual consumers cannot unilaterally value economic goods differently from their valuation in the market. Valuation is necessarily a collective not an individual activity. A market may therefore be described, amongst other things, as a collective mechanism for valuing goods. If a consumer values a good he buys at more than its value in the market, he will simply buy more, and if he values a good at less than the market he will simply buy less or not buy that good at all. Furthermore, market value is certainly the way consumers express value when they discuss the value of goods. Except perhaps for students of economics, it is unheard of for consumers to discuss value in terms of some distinct, vague and un-quantifiable concept of utility.

Many consumers value some goods they own at more than their market value. When they do, they know this is irrational and uneconomic. This is what we mean by sentimental value. The implication is that such goods are not economic goods and their owners not behaving as 'economic men' with respect to those goods. All economic goods, on the other hand, that are purchased and consumed are valued as the market values them. We may therefore characterise this theory as 'The Market Theory of Value' since in this system all economic goods are assumed to be valued, by all consumers, as the market values them.

It now becomes evident why the common practice of substituting real values as proxy variables for supposedly un-measurable utilities in econometric regression analysis works so well. Utility is nothing more than real value. The fact that utility is measured in terms of money also makes comparisons between the utility derived from consumption of each good and available financial resources quite simple. It also becomes evident how consumers can simply 'optimise' the consumption of each of the goods they consume to maximise their utility. If consumers were actually to formally maximise their utility the optimum condition would arise at the point where the ratio of marginal utility to the market price is equal for all the goods they consume. But since all consumers value all the goods they consume at their market price then this ratio is always equal to 1 irrespective of the quantity of each good a consumer may buy.   Optimisation therefore could not be simpler.   Each consumer just follows his subjective preferences and priorities, allocates his resources as he subjectively prefers and buys whatever he pleases.   If this were not simple enough this procedure is capable or further practical simplification. Formal optimisation is generally unnecessary since consumers can simply approximate optimisation by paying particular attention to high value goods

Resale Markets

Typically the market price of second hand goods is low compared with their original purchase price except for those goods that attract speculation. This partly reflects wear and tear reducing the 'intrinsic' value of goods and partly a widespread consumer preference for new goods.

This condition is aggravated where there is no organised resale market or where the resale market is a 'marginal' trade.   More expensive goods and goods that have a long useful life are less prone to these effects than other goods. Prime examples of these are houses and cars. As a consequence, large organised resale markets exist for these goods.

Generally, consumers value purchases at their original purchase price, the opportunity cost of obtaining them.   Where goods have a long life and widespread organised resale markets exist however, consumers may become aware of the current market price of goods they own. Houses and cars for example are constantly advertised for sale in many newspapers as well as in publications specifically intended to assist the resale market.

For rare goods like valuable works of art or antiques, particularly where they were inherited rather than purchased, knowledge of their current value may induce the owner to sell. When they do not sell, even though they are aware of the current market price, their preference for the good effectively means that the value they place on the good reflects the current market price, or opportunity cost of retaining the good rather than the original price or value, or the goods hold a higher sentimental value for them.

Comparison of Optimisation Procedures Under Ordinal Utility Theory & The New Theory of Value

As we have seen the way a consumer allocates his resources under the new theory of value is intrinsically simpler than that assumed in ordinal utility theory.

Even consumers with quite modest incomes will typically consume hundreds of differentiable goods and services. Those with higher levels of income may consume thousands of different goods. If an individual consumes n different goods, ordinal utility theory requires they optimise a system of $^nC_2$ two-dimensional indifference maps.

Now
$$^nC_2 = \frac{n!}{(n-2)!\,2!}$$

where $^nC_2$ is the number of possible combinations of any 2 goods from n goods

and n! Is factorial n

If an individual consumes 1001 different goods, and we interpret this literally and not figuratively, this requires he optimises 500,500 two-dimensional indifference maps. It is doubtful that even the most mathematically gifted could formally optimise such a system unaided. Even allocating the budget constraints, among each of 500,500 pairs of goods, is a much more complicated problem (involving the solution of a system of 500,500 simultaneous equations) than resource allocation under the new theory of value. The complete optimisation process under ordinal utility theory may also be iterative further increasing the complexity of the problem. Consumers clearly do not maximise their utility function in the way that ordinal utility theory says they should.

Furthermore, ordinal utility theory only allows an economist to avoid calculating utility, not the consumer. The consumer must be able to evaluate utility, at least relatively, in order to define his $^nC_2$ indifference maps even before he can start to optimise his behaviour. Just remembering 500,500 indifference maps would be a serious problem. Ordinal utility theory proposes a ridiculously complex optimisation problem. It therefore cannot be correct.

By comparison the method suggested for optimising the cardinal utility system proposed in this essay is surprisingly simple. The fact that most consumers are more or less able to balance their budgets demonstrates that most people can allocate their resources to competing uses reasonably effectively. This is all the new theory of value requires, an approximately balanced budget (including borrowing) and a subjective allocation of that budget to each good consumed.

Chapter 4. A New Theory of Demand

Deriving the Demand Function for an Individual

From equation 3 and 11 we can derive an individual's short-term demand function reflecting his own preference and priority for consumption of a good in two forms.

The 'conventional' form of an individual's short term demand function is given by:

$$P_x = (\alpha_x y e^{-y/\lambda} + \beta_x)/ q_x \qquad \text{Equation 15}$$

and the 'natural' form of an individual's short term demand function is given by:

$$q_x = (\alpha_x y e^{-y/\lambda} + \beta_x)/ P_x \qquad \text{Equation 16}$$

Equation 15 is described as the conventional form of the demand function because when represented on the traditional supply and demand graph P_x is shown on the Y axis, which conventionally carries the dependent variable. P_x is also treated as a dependent variable for all x in General Equilibrium analysis. It is therefore treated in that way here though price is often thought of as the independent variable in demand analysis. We will return to consider the likelihood that equation 15 is the 'true' form of the short-term demand function later.

Equation 16 is described as the natural form of the demand function because it enables us to understand how an individual allocates his resources: assets or initial endowment, income and borrowing to competing uses. The quantity demanded q_x is simply that quantity that an individual consumer can afford to buy at the prevailing market price, with the allocation of his resources he chooses to make to good x, in accordance with his preference α_x and priority β_x.

To extend our analysis of consumer choice, looking at demand in this way enables us to understand consumers search behaviour. If total expenditure on a particular good represents a trivial proportion of resources, the consequence of an error in maximising utility by a mis-purchase will be slight, a mis-purchase being defined as a purchase that a consumer later regrets because he subsequently finds he values the good at less than its market value. A consumer will therefore spend very little time in making his choice of trivial purchases. As the consequences become greater, the care a consumer needs to take must become greater. This is why routine purchases in a supermarket may take only a few seconds to select. The purchase of a suit of clothes or a pair of shoes involving visits to several shops and trying on various items may take perhaps an hour or so. For more substantial purchases like a holiday or a car a consumer may take several weeks comparing alternatives before he makes his choice. When buying a house, the search could take even longer perhaps several months. A consumer's transaction cost is therefore a behavioural variable, a function of the consequence of error. The already very simple optimisation process of the new theory of value is therefore capable of further practical simplification where the consequences of error are slight. Incidentally this also explains why the poor on tight budgets take more care over all their purchases than the more prosperous.

We have all experienced important buying decisions where the price of the good we most desire (perhaps a particular new car) is slightly more than we intended to pay (our 'optimal' allocation of resources). We then have to make a choice between a less desired good that nevertheless meets our basic needs and our most desired good at a slightly higher price than we intended. Usually this involves much heart searching whatever the final choice. This behaviour is perhaps as near as a consumer will ever come to formal optimisation.

Further assumptions of neoclassical consumer theory now need to be jettisoned. Firstly, information is not perfect or free. Consumers seek out information in proportion to the consequence of error. In a perfectly competitive market there will be an infinite number of vendors. To consider all of these will require an infinite search, taking an infinite period of time. No consumer can afford to squander time on protracted searches, so consumer choice must involve a simple satisficing routine that elaborates the search in proportion to increasing consequence of error. Incidentally a similar limited search routine will apply to firms when seeking to benchmark their prices against their competitors.

The selection of trivial purchases could therefore be idiosyncratic. Indeed, more prosperous consumers may even suspend optimisation altogether for trivial purchases and use the opposite rule of paying whatever it costs to obtain what they seek. Such search routines are more consistent with an institutionalist approach to economics than a neoclassical approach.

Deriving the Market Demand Curve

The market demand curve for a good in a competitive market can be derived in the natural form of equation 16 by the summation of all the individual demand curves of all consumers of that good in the economy since consumers are almost invariably price takers and therefore the price paid by each consumer is common to all their individual demand curves.

Therefore, from equation 16:

$$D_x = (\sum_{1}^{N} (\alpha_x y e^{-y/\lambda} + \beta_x))/ P_x \qquad \text{Equation 17}$$

Where D_x is the market demand for good x
P_x is the market price of good x
and N is the population within the economy

Note that some consumers' resources, preferences and priorities for good x may be such that they consume none at all.

Simplifying equation 17 by representing the sum of individual preferences and priorities across the population using the results of equation 6 gives:

$$D_x = (a_x Ye^{-Y/N\lambda} + b_x)/ P_x \qquad \text{Equation 18}$$

or

$$P_x = (a_x Ye^{-Y/N\lambda} + b_x)/ D_x \qquad \text{Equation 19}$$

when expressed in conventional form.

Equation 19 could be characterised as the short term competitive demand function since a market can only clear when the total sales value of a good brought to market $P_x D_x$ is sold and equal to the total expenditure on that good V_x. It follows that equation 19 must hold in the short term for a competitive market to clear since α_x and β_x and hence a_x and b_x are behavioural variable that are unlikely to change in the short term.

The term competitive market is generally taken to imply a market in which firms as well as consumers are price takers. It is also commonly used to differentiate competing firms from monopolists or oligopolists. It is not used in that way here, since this work is solely concerned with demand not supply. The distinction intended is between a competitive market, to which the demand function derived here applies, and a speculative market that is subject to a different demand function. The term 'competitive' as used in this context therefore applies only to the normal economising behaviour of consumers or other buyers as prices rise. Thus, competitive markets are distinguished from markets in which, consumers (or other buyer) fail to economise as prices rise (for example in speculative and **Giffen** good markets). Such an approach to the definition of a competitive market bears greater similarity to real markets in a competitive economy in which a mix of monopolists, oligopolists and competitive firms (in the conventional sense) all trade together. When we wish to refer to a competitive firm or market in a conventional sense, we will use the term 'perfectly competitive'.

Using this definition, oligopoly and monopoly trading are forms of competitive trading, because they both depend on the same demand function. The latter represents the limiting condition for a competitive market. This may seem to be rather like the oriental notion of the sound of one hand clapping but in fact it is not. In this system it is solely the number and behaviour of consumers or other buyers that determines whether a market is competitive, not the number or behaviour of firms or other vendors. Figure 7 shows the market demand function for a competitive market.

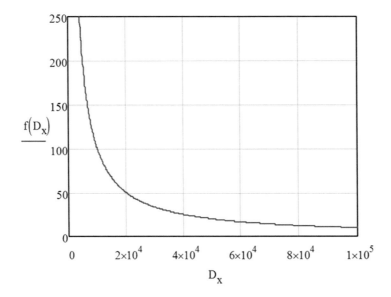

Figure 7. Market Demand Function for a Competitive Good

Figure 7 certainly shows the expected form of the competitive demand function, is similar to demand functions shown in 19th century texts such as (Marshal, 1890), and is preferable to the declining linear function show in many modern texts such as (Mackintosh el al, 1996). Why we adopted this approach will become apparent when we return to the question of value. For competitive goods we can derive the slope of the market demand function and the market price elasticity of demand as follows.

Firstly, the condition:

$$\frac{dP_x}{dD_x} < 0$$

always applies to competitive goods.

Secondly since $V_x = P_x D_x$

the competitive demand function can be represented as:

$$P_x = \frac{V_x}{D_x}$$

a form of equation 19. Therefore:

$$\frac{dP_x}{dD_x} = (D_x\frac{dV_x}{dD_x} - V_x)/D_x^2 \qquad \text{Equation 20}$$

substituting for V_x $\frac{dP_x}{dD_x} = \frac{dV_x}{D_x dD_x} - \frac{P_x}{D_x}$

Now at any point in time with any level of income and any state of preference and priority V_x can be assumed to be constant.

Therefore
$$\frac{dV_x}{dD_x} = 0$$

and
$$\frac{dP_x}{dD_x} = -\frac{P_x}{D_x} \qquad \text{Equation 21}$$

Now the price elasticity of demand is given by:

$$\mathcal{E}_{Px} = \frac{P_x dD_x}{D_x dP_x}$$

therefore
$$\mathcal{E}_{Px} = -1 \qquad \text{Equation 22}$$

Thus, contrary to neoclassical orthodoxy, the price elasticity of demand is always equal to -1 for the short-term demand function of a competitive good. This approach is used because it is easier to differentiate the dependent variable in an equation with respect to the independent variable than vice versa, and because in the conventional form of the demand function price is assumed to be the 'dependent' variable, since price appears on the y axis of a conventional demand function graph.

Most economists would accept that equation 19 represents one form of the competitive demand function but few, if any, would accept that this is the sole form. It seems unlikely however that two similar but different curves involving the three variable P_x, V_x and D_x can exist. That equation 19 is the sole form of the short-term competitive demand function is one of the key conjectures of the new theory of demand. The theory developed here would largely still work were this not so, though the mathematics would be more complicated. While a theory that did not insist on the generality of equation 19 would be easier to sell, it would not be correct so we must take up the challenge now and try to show why the conjecture of the theory is correct. We cannot prove this conjecture, though it would be easy to disprove. All that is needed is to find a significant set of goods (not just a few isolated examples) in a competitive market (as defined here) in which a wide variety of prices and levels of demand arise (even in the short term) that do not have a demand functions following equation 19.

The form of demand function derived from econometric investigations will necessarily involve collecting data over time rather than at an instant in time. Large changes in prices in competitive markets (as distinct from speculative markets) generally only arise over a long period of time rather than rapidly over a short period. Econometric investigation will generally therefore only ever deduce a long or at best medium-term demand function not the short-term demand function proposed here.

Similarly the theoretical derivation of an individual's demand function from an indifference map implies prices at different levels of demand apply at different points in time and not at an instant. This is because different prices are derived at different levels of income. Though occasionally an individual may suddenly achieve a much greater income by inheritance or by winning a national lottery, increased income generally only arises over time as a person's career advances. In addition theoretical demand curves (derived from indifference maps) that show elastic or inelastic demand functions also imply a change of preference. Falling preference as incomes rise would produce an inelastic long-term demand function and rising preference as incomes rise would produce an elastic long-term demand function. Because preference is learned behaviour (arising from trial and error) a change of preference must necessarily only ever occurs over time, not at an instant in time. This idea that time affects the form of the demand function is by no means new, For example (Alfred Marshal, 1890) discusses the difficulty that gathering data over time causes in ascertaining the form of the demand function.

Many economists will be uncomfortable with this formulation however, because it appears to be an un-testable hypothesis. While testing this theory is more difficult and such tests cannot be applied to such a wide variety of goods, the theory is nevertheless testable for any competitive goods for which the production process is uncertain and output fluctuates widely week by week. The attention of econometricians is directed to such markets that should provide excellent opportunities to test the theory.

We can give two very different examples of this, soft fruit and computer processor chips (large scale integrated circuits). Note that, the market structures of these two product groups differ greatly. The former are grown by small-scale producers in many regions of the world, approximating perfect competition, whereas the latter are manufactured under oligopoly or even monopoly conditions in plants of great industrial concentration. If the short-term demand function could be shown to be consistent with conditions in such markedly different markets this would be compelling evidence for the correctness of the theory.

The yield of soft fruit depends, amongst other things, on the amount of rainfall over the growing season. Generally (but see below), higher levels of rainfall produce higher yields of soft fruit. If therefore the amount of rainfall over the growing season fluctuates widely up and down week by week, the yield of soft fruit will vary substantially week by week and therefore the price will vary substantially even in the short term.

Provided rainfall is not torrential late in the growing season, the chance fluctuations in supply will result in significant short-term fluctuations in the spot price of soft fruit. These conditions are an excellent framework in which to test the theoretical short-term demand function. Torrential rainfall late in the growing season however, damages the fruit and causes a catastrophic failure of the harvest and a speculative market in soft fruit to develop. These conditions would be unsuitable for testing the theory.

Similarly, though for different reasons, the yield of the fastest types of computer processor chips is very low, typically just a few percent of total production. It can transpire that for weeks at a time no acceptable chips are produced. The Japanese describe such a production system as an incapable process because it is incompatible with their methodology of total quality management. They nevertheless persist with the recalcitrant technology and produce a substantial proportion of such chips.

If the yield of acceptable chips fluctuated up and down week by week this too would cause a fluctuation in the spot price of a particular processor chip. Provided there was no catastrophic failure of production this too provides an excellent framework for testing the theory. Should a serious fire occur at one of the key production plants, or a very long period of zero yield at a sole producers plant, this would once again cause a speculative market to develop. These conditions are unsuitable for testing the theory.

While we cannot prove equation 19 we can show why it is unlikely that a competitive demand function can take any other form in the short term. Equation 19 could be described as the fundamental competitive demand function since it has unit elasticity (actually −1). One might expect that any other competitive demand function would be related to equation 19 in some way. One possible way of generalising equation 19 is given below:

$$P_x = A_x \cdot \frac{V_x}{D_x} + B_x \qquad \text{Equation 23}$$

Where A_x and B_x are two parameters or arbitrary constants unrelated to similarly labelled terms used in the aggregate consumption function.

This equation defines a set of 'broadly inverse' functions that bear a similar relationship to the 'fundamental' inverse demand function of equation 19 that the set of all linear functions bear to

$$y = x.$$

Differentiating this 'generalised' expression we obtain:

$$\frac{dP_x}{dD_x} = -\frac{A_x \cdot V_x}{D_x^2}$$

Now

$$\varepsilon_{px} = \frac{P_x}{D_x} \frac{dD_x}{dP_x}$$

therefore
$$A_x = \frac{-1}{\varepsilon_{px}}$$

and
$$P_x = \frac{-1}{\varepsilon_{px}} \frac{V_x}{D_x} + B_x$$

This equation seems to have great potential as a general form of demand function because it is related to equation 19 and includes elasticity, which it is generally held can take any value in the open interval:

$$0 > \varepsilon_{px} > -\infty$$

If this were so the demand function could take any form from nearly horizontal to nearly vertical. The graph below shows the form of demand function that such an equation implies for various price elasticities:

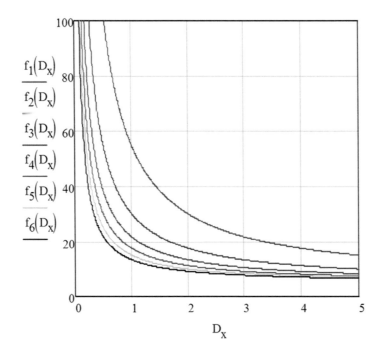

Figure 8. Possible Demand Functions of Decreasing Price Elasticity

The red line has –0.2 price elasticity, the torques line unit – 1 elasticity and the black line an elasticity of –1.2 Unfortunately equations such as these exist only in the minds of economists not in the real world since:

$$B_x = P_x \cdot \left(1 + \frac{1}{\varepsilon_{px}} \right)$$

$$P_x = \frac{-1}{\varepsilon_{px}} \frac{V_x}{D_x} + P_x \cdot \left(1 + \frac{1}{\varepsilon_{px}} \right)$$

$$\frac{-1}{\varepsilon_{px}} P_x = \frac{-1}{\varepsilon_{px}} \frac{V_x}{D_x}$$

V_x must always be equal to $P_x \cdot D_x$ since in a competitive market the value of consumption (or total expenditure if you prefer) must always be equal to the product of the competitive equilibrium price and the effective demand quantity for the market to clear, irrespective of what the level of demand, supply or price may be from time to time. This applies both to each individual consumer and to the market as a whole.

Therefore:

$$\varepsilon_{px} = -1$$

$$A_x = 1$$

$$B_x = 0$$

The form of equation 23 therefore reverts to that of equation 19. Similarly, we could adopt a more general form of demand function such as that shown below.

$$P_x = \frac{V_x}{D_x} \cdot f(D_x)$$

This equation is in fact entirely general since $f(D_x)$ can contain any number of terms, some or all of which could contain:

$$\frac{V_x}{D_x} \quad \text{or} \quad -\frac{V_x}{D_x}$$

but

$$V_x = P_x \cdot D_x$$

and therefore:

$$f(D_x) = 1$$

It follows that if you accept that V_x must always be equal to $P_x \cdot D_x$, in the short term, for the market to clear, then the only form the competitive demand function can take in the short term is that of equation 19. We will provide further evidence for this when we investigate the conditions for general equilibrium.

Therefore, contrary to neoclassical orthodoxy, the price elasticity of demand for a competitive good in the short term is always equal to -1. It has to be that way otherwise the demand function will not represent scarcity correctly and consumers would not economise in the use of such goods correctly. For competitive goods price has to be inversely proportional to demand in the short term and vice versa for the economy to work. A constant price elasticity of demand also explains why consumers are 'indifferent' to changes in the quantity of a good that their resource allocation enables them to buy and why optimisation under the new theory of value is so simple. Consumers are indifferent because at a point in time the value consumed must always be the same irrespective of the equilibrium level of price.

So why have economists held for so long that elasticity could take any of the values given above. The misconception arises from an illusion caused by linear regression. Note this is the second example of the unsuitability of linear regression (at least in the form it has customarily been used) as a means of econometric investigation. Linear regression is the standard empirical method used to model demand functions. For estimating the slope and position of a particular region of a demand function of interest this is a reasonable enough procedure. It does however create an illusion of varying elasticity since the elasticity of a linear demand function will vary over its entire length. The illusion of varying price elasticity is intensified when such models are applied to goods in short supply such as precious metals or precious stones or to 'goods' where the supply is restricted such as illegal heroin. For such goods the operative part of the demand curve is that part which is asymptotic to the price axis. A linear model would represent this as a nearly vertical line. Similarly for those goods in plentiful supply the operative part of the demand curve would be that part which is asymptotic to the demand axis. A linear model would represent this as a nearly horizontal line. Our crude models therefore obscure the true relationship by giving the impression of almost perfect inelasticity or almost perfect elasticity. The graph below shows a demand function of the form of equation 19 and two linear functions fitted to it using linear regression.

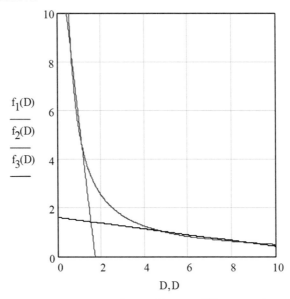

Figure 9. Linear Regression Over Two Different Regions of Demand

Both of the two linear functions achieve very high correlation coefficients for data derived from the demand function, $- 0.97$ in each case. The linear functions are indeed very good models of the demand function over the regions for which they were defined. An econometrician who obtained such well-fitting models would think he had done a good job. In fact, both functions are a gross distortion of the function they try to represent and would be very poor predictors of the behaviour of the demand function for points well outside the range of the data sets used to define the models.

While some evidence has been presented to indicate that the conjecture of the new theory of demand concerning the sole form of the short term competitive demand function may hold and some reasons why the form of the function may have been misconceived in the past, none of this is entirely convincing. We can however formulate the conjecture of the new theory of demand as a testable hypothesis. One problem with this however is that the conditioning of non-linear regression is far worse than that of linear regression. Thus, even though linear models may be much less representative they are far less sensitive to errors in the data than non-linear models. This is one reason why linear regression is used so widely. The other main reasons are that it is much easier to use and requires no a priori knowledge of the form of the function involved. Despite conditioning difficulties however we should be able to state a testable hypothesis in the following form:

For any wide range ($P_{Max} > 10.P_{Min}$ AND $D_{Max} > 10.D_{Min}$) of homogeneous price and demand data collected over a short period of time, from a competitive market in which the level of output is subject to chance, and can fluctuate widely even in the short term, if equation 19 is fitted to the data using non-linear regression, (or using linear regression, a single independent explanatory variable, calculated from V/D, and an intercept constrained to zero) the resulting R^2 value will at worst not be statistically significantly lower, and at best will be statistically significantly higher, than that derived from any linear function fitted to the same data using linear regression, untransformed (or recalculated) data and an unconstrained intercept. The gauntlet is thrown down, let econometricians do their worst.

Estimating a Firm's Demand Function

If a firm knows the total value v_x of its sales of a particular good x over a short period of time and x is a competitive good, the firm can plot the demand curve for x prevailing at that time since:

$$P_x = \frac{v_x}{q_x}$$

If output is expressed in sales value per week or month then demand is expressed as units per week or month. Note for firms with multiple products this holds for each product separately and not necessarily for sales in aggregate. Even different variants of the same good may in fact have different demand curves and may need to be treated separately. The new theory of value and demand can therefore provide the benefit of greater precision in calculating the effect of a change in price on sales for most firms. Hitherto few firms knew the shape of their demand curve. Now most firms do.

The same approach could be applied to annual sales and annual demand though it is much less likely that conditions will remain stable for a year at a time in many markets.

Chapter 5. The Competitive Market Equilibrium

We will start with a typical neoclassical analysis of pricing in a competitive market. Figure 10 below shows the demand function, which economists also describe as the average revenue function and a typical so called average cost function. Though these two functions are described as 'average' functions, they are in fact the unit price function of demand and the unit cost function of supply at any volume. The cost function used here is similar to the concept of cost widely used in much commercial practice except that 'exceptional costs', discussed below, are collected by firms as additional expenditure in their accounts and not modelled by an approximation as proposed below.

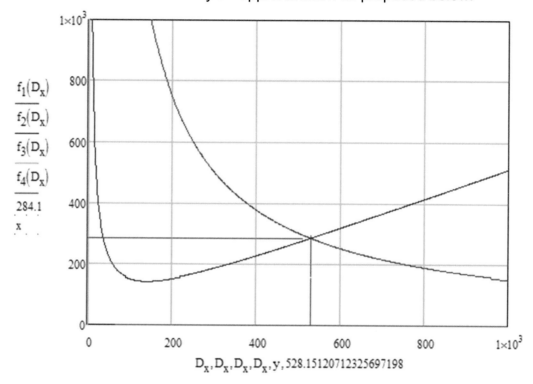

Figure 10 The Competitive Equilibrium of Demand & Supply

Equilibrium in this market is assumed to arise in Neoclassical Economics at the intersection of the marginal revenue and marginal cost functions. Marginal revenue is defined as the income arising from one more unit of sales and marginal cost as the expenditure incurred to produce one more unit of production. Mathematically these functions can be represented as the first derivative of total revenue and the first derivative of total cost respectively.

One difficulty with this approach is that there is no generally accepted or universally applicable formula for marginal cost. The form of the marginal cost function is a supply side issue and a matter of great debate in economics. For the time being I will assume the form of this function is as hitherto thought. We will see later that this assumption is problematic. Broadly two forms of 'marginal' cost curves are thought to exist, U shaped and L shaped. The latter implies more or less constant marginal cost over a wide range of output. Unfortunately, far fewer firms are thought to have L shaped marginal cost curves than U shaped since these mostly apply to large-scale producers. Furthermore even where it does apply the L shaped form of cost curve is particularly associated with the long-term, not the short-term cost function. Almost all firms would be expected to have U shaped short-term cost curves because increasing output in the short term incurs additional costs such as overtime payments and shift premiums.

Short-term conditions are used throughout this essay under the presumption that market conditions will normally be disturbed well before long-term equilibrium conditions are established. This is the more general case. If long-term equilibrium conditions were ever established in a particular market this would be the exception and not the norm.

We can make what is perhaps a more realistic assumption (based on the 'analogous' accounting concept of direct cost) that short-term unit costs follow an equation of the form given below or a similar form.

$$C_D = \frac{C_F}{q_x} + C_V + C_E.q_x \qquad \text{Equation 24}$$

an example of the, so-called, average cost function where:

C_D direct cost is the unit cost of good x
q_x is the quantity of good x produced
C_F is fixed or capital cost
C_V is variable cost
and $\quad C_E$ are exceptional costs such as overtime payments, shift premiums, short term purchasing premiums or discounts lost.

These exceptional costs would in general be unique to each good, firm and time and be discontinuous. The only practical way to model such costs however is to assume they are at least linear (an approximation to a series of small step function increases in exceptional costs) or of higher order.

Irrespective of whether this formula is generally accepted, all will agree it is a formula for a U-shaped cost curve. An example of this function is plotted above and below.

The graph in figure 12 below is produced by MATHCAD using the demand function derived in equation 19 above to plot the red line, giving the unit revenue function in the form:

$$f_1(D_x) := \frac{V_x}{D_x}$$

with parameter V_x = £150,000

and the unit cost function of equation 24 above to plot the blue line in the form:

$$f_2(D_x) := \frac{C_F}{D_x} + C_V + C_E \cdot D_x$$

Substituting D_x for q_x with parameters $C_F = £100,000$, $C_V = £1$ and $C_E = £0.5$

The first problem with the conventional economic analysis of marginal revenue and marginal cost is that total revenue derived from equation 6 above is constant in the short term because the preference and priority of all consumers is constant in the short term. This must be so since preference and priority are learned behaviour that can only change in the medium or long term not instantaneously in the short term. Marginal revenue derived from the first derivative of equation 11 above is therefore always equal to zero in the short term.

The second problem is that if we derive total cost C_T from equation 24 above as the product of D_x and unit cost C_D the first derivative of C_T evaluates to:

$$\frac{dC_T}{dD_x} = C_V + 2C_E D_x$$

that can only be equal to zero when $D_x < 0$ and outside the domain of D_x because you cannot have negative demand. This result could explain why firms find it so difficult to calculate their marginal cost.

Practically therefore we must base our definition of equilibrium on unit cost and not the first derivative of total cost and on unit revenue not the first derivative of total revenue as defined above. Since both unit revenue and unit cost are downward sloping curves, at least initially. The slope of both curves is negative and only appears in the second and fourth quadrants of D_x in any graph showing demand and supply in a competitive market such as that shown in figure 11 below.

The third problem is that if there is a term in D_x or of higher order in the unit cost function the equation

$$\frac{dC_D}{dD_x} = \frac{dV_x}{dD_x}$$

relating the first derivative of unit cost and the first derivative of average revenue resolves to a quadratic equation in D_x or an equation of higher order both with no real roots.

Any such form of optimum economic solution cannot therefore be correct and equilibrium can only occur whenever demand is equal to supply irrespective of the inequality of marginal revenue and marginal cost. Figure 12 shows the first and fourth quadrants of a graph similar to figure 11, including the marginal unit revenue function or rate of change of price in green and the marginal unit cost function or rate of change of cost in pink, both appearing here in the fourth quadrant illustrating this point.

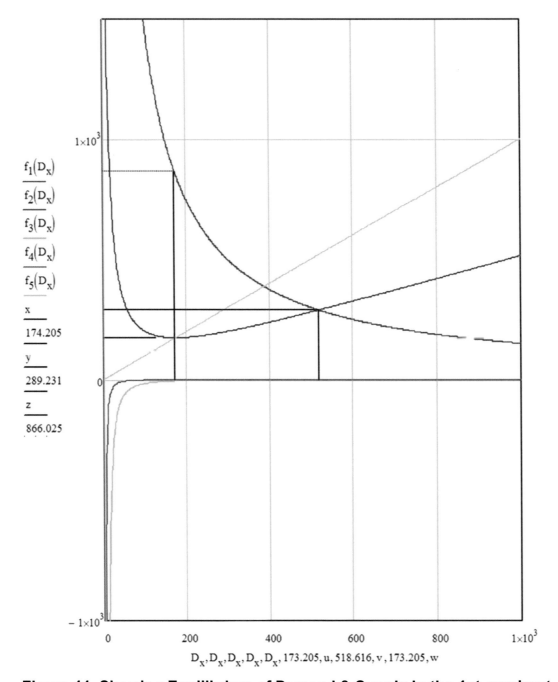

Figure 11. Showing Equilibrium of Demand & Supply in the 1st quadrant, and their first derivative functions in the 4th quadrant

Figure 11 also shows $f_5(D_x)$ the first derivative of total cost plotted in light blue derived above which will intersect $f_2(D_x)$ at the turning point of C_D where unit cost is at its minimum and the first derivative of unit cost is equal to zero. Note the intercept of $f_5(D_x)$ with the y access is at C_E not at zero. We will explore this issue further in the next chapter when we consider 'Monopoly Pricing'.

This intersection also provides a means of calculating the optimum condition in a competitive market where fixed cost in the whole market will naturally increase by investment as a result of competition to the point where a simultaneous intersection of $f_1(D_x)$, $f_2(D_x)$ and $f_5(D_x)$ occurs as shown in figure 12 below.

This condition is determined by solving the equation:

$$f_1(D_x) = f_5(D_x)$$

$$\frac{V_x}{D_x} = C_v + 2C_E D_x$$

for D_x with parameters V_x, C_v and C_E with the values used in our example above giving the result

$$D_x = 386.799$$

to 3 decimal places and thence solving the equation:

$$f_2(D_x) = f_5(D_x)$$

$$\frac{C_F}{D_x} + C_v + C_E D_x = C_v + 2C_E D_x$$

for C_F with parameters C_v and C_E as above, where D_x has a value of 386.799 its value at the intersection of $f_1(D_x)$ and $f_5(D_x)$ giving the result:

$$C_F = £74,806.73$$

to 2 decimal places. This condition gives the optimum state where market demand is produced at minimum unit cost a natural equilibrium condition for a competitive market as a whole. Note the effect on $f_4(D_x)$, which is now much closer to $f_5(D_x)$.

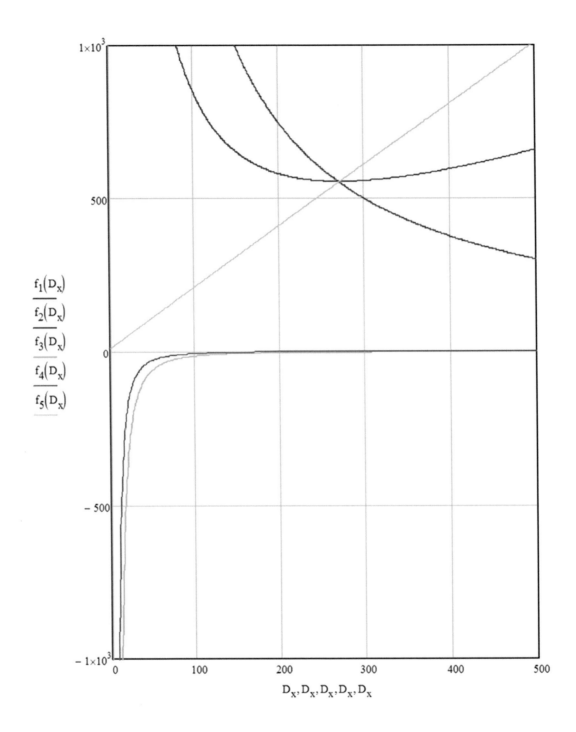

Figure 12 Showing the optimum condition for the consumer in a competitive market

Chapter 6. The Conditions for Monopoly Pricing

You may be wondering why this chapter concerned with monopoly pricing (a supply side issue) appears in a theory of value and demand. There are three reasons:

1. Price is a demand side issue and monopoly pricing raises important issues concerning value.
2. We can derive a general pricing solution under all conditions of market structure from a simplified monopoly pricing solution.
3. We can derive the speculative demand function from the same equation used to derive the simplified monopoly pricing solution.

The conditions for monopoly pricing in a natural or true monopoly or under monopolistic competition between sharply differentiated branded goods can also be derived using a typical neoclassical analysis. Figure 13 shows the short-term demand function or average revenue function and the short-term direct cost function and possible conditions for monopoly pricing.

Though the demand function is often referred to as the average revenue function it is not however strictly an average function at all but a function of total revenue per unit at any level of demand. See figure 21 below, that shows the true average revenue function together with the two functions of figure 10 and 13 to illustrate the difference. Average revenue is in fact calculated by integrating the short term demand function, in principle between zero and D_x, and dividing by D_x. The same principle also applies to calculating average cost. We will however continue to use the terms average cost and average revenue in the way other economists use these terms so as to avoid confusion.

The over simplified analysis used to define monopoly conditions, such as the graph shown in my own basic economics textbook on page 193 (M. Mackintosh et al, 1996) and many other similar works, is impossible and cannot apply because if the demand function is linear with negative slope, marginal revenue is constant and negative and can therefore never intersect with any long run marginal cost function which is also constant but positive or zero. These two curves can therefore never intersect.

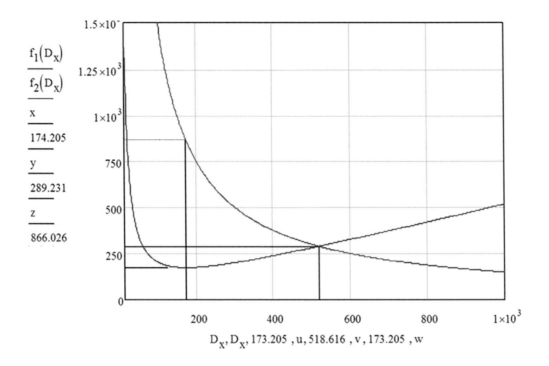

Figure 13. Conditions for Monopoly Pricing

Figure 13 is drawn assuming the short-term demand function derived above and the short-term direct cost per unit function of equation 24 above with parameters V_x = £150,000, C_F = 15,000, C_V = 1, C_E = 0.5

The monopolists' profit is represented by: the area bounded by the y axis and the vertical black line on the left (effective demand) of 173.205 and the x axis and the horizontal black line at the top, giving a monopoly price of £866.026 minus the area bounded by the y axis and the vertical black line (effective demand) of 173.205 and the x axis and horizontal black line at the bottom (direct cost) of £174.205 to 3 decimal places. The values of these variables are shown on each axis. In this example monopoly profit is approximately £119,827 using the method of calculation derived below. The equilibrium conditions in a conventionally defined perfectly competitive market subject to the same demand and cost functions are also shown by the vertical and horizontal black lines at the intersection of the demand and direct cost functions. In this example the short-term competitive equilibrium point actually results in an aggregate loss (for the market as a whole) of approximately £11,526 as shown in figure 21 below. This illustrates and reinforces the contention that there is no such thing as a normal profit under a short-term equilibrium.

Firstly, I will derive a simplified solution, which produces an approximation to the optimum monopoly profit and can subsequently be used to derive the speculative demand function.

Using this approach, the monopolist's profit Z_x for good x is assumed to be:

$$Z_x = P_x \cdot D_x - \int_1^{D_x} C_D \, dD_x$$

therefore $\quad\quad\quad \dfrac{dZ_x}{dD_x} = P_x + D_x\dfrac{dP_x}{dD_x} - C_D \quad\quad\quad$ Equation 25

now under the competitive demand function:

$$D_x = \frac{V_x}{P_x}$$

Thus $\quad\quad\quad \dfrac{dZ_x}{dD_x} = P_x + \dfrac{V_x dP_x}{P_x dD_x} - C_D \quad\quad\quad$ to a first approximation

If a monopolist seeks to maximise profits then:

$$\frac{dZ_x}{dD_x} = 0$$

thus $\quad\quad\quad 0 = P_x + \dfrac{V_x dP_x}{P_x dD_x} - C_D \quad\quad\quad$ Equation 26

therefore $\quad\quad\quad P_x^2 - C_D P_x + V_x\dfrac{dP_x}{dD_x} = 0$

and $\quad\quad\quad P_x = \dfrac{C_D \pm \sqrt{(C_D^2 - 4 V_x\dfrac{dP_x}{dD_x})}}{2} \quad\quad\quad$ Equation 27

where $\quad\quad\quad Z_x$ is profit derived from good x

and $\quad\quad\quad C_D$ is the direct cost of good x

and since $\quad\quad\quad D_x = \dfrac{V_x}{P_x}$

then $\quad\quad\quad D_x = \dfrac{2V_x}{C_D \pm \sqrt{(C_D^2 - 4 V_x\dfrac{dP_x}{dD_x})}} \quad\quad\quad$ Equation 28

The position of the two sets of black lines in figure 13, are consistent with these results. Readers may be concerned that this quadratic solution contains a differential coefficient. $\dfrac{dP_x}{dD_x}$ is the slope of the demand function. Estimating the slope of the demand function in a particular region is a common econometric procedure. Inserting a parameter for $\dfrac{dP_x}{dD_x}$ using a local estimated value for the slope of the demand function would reduce the quadratic solution to purely algebraic form. The same treatment of differential coefficients is used in Newton's equations of motion. It would be difficult to find a more respectable precedent. Alternatively the first derivative of price with respect to demand, in a competitive market, can be calculated from equation 21.

It appears therefore that monopoly prices are bi-stable. In fact the lower level of price in the simplified monopoly pricing solution is either negative or unreal or gives the minimum price to avoid making a loss and the higher price is that which maximizes profit. However, see below for a more refined treatment of this analysis.

Now to return to the analysis of monopoly prices and profits, this part of my analysis began with a quest to find the optimum level of output for a monopolist that yields the greatest profit. The monopolist's profit Z_x for good x can be written as:

$$Z_x = V_x - C_T$$

where $\qquad Z_x$ is profit derived from good x

which can be expanded to $\qquad Z_x = (P_x - C_D)D_x$

This can be written very elegantly in functional notation as:

$$f_5(D_x) = (f_1(D_x) - f_2(D_x))D_x \qquad \text{Equation 29}$$

where $\qquad f_5(D_x) = Z_x$

$\qquad f_1(D_x) = P_x$ from a form of equation 19

and $\qquad f_2(D_x) = C_D$ from equation 24

then the optimum level of profit can be found by first solving the equation

$$\frac{d(f_5(D_x))}{d(D_x)} = 0 \text{ for } D_x$$

hence $\qquad \dfrac{d[(f_1(D_x) - f_2(D_x))D_x]}{d(D_x)} = 0$

Though this equation looks simple it's actually quite complicated resulting in a cubic function in D_x which MATHCAD can solve as follows:

$$\left[(f_1(D_x) - f_2(D_x)) \cdot D_x\right] = \left(\frac{V_x - C_F}{D_x} - C_E D_x - C_V\right)D_x$$

$$\left[f_1\left(D_x\right) - f_2\left(D_x\right)D_x \right] = D_x \cdot \left(\frac{135000}{D_x} - 0.5.D_x - 1 \right)$$

$$\frac{d}{dD_x}\left[\left(f_1\left(D_x\right) - f_2\left(D_x\right)\right) \cdot D_x \right] \rightarrow \frac{135000}{D_x} - D_x \cdot \left(\frac{135000}{D_x^2} + 0.5 \right) + -0.5.D_x - 1$$

$$\frac{135000}{D_x} - D_x \cdot \left(\frac{135000}{D_x^2} + 0.5 \right) + -0.5.D_x - 1 = 0$$

$$135000D_x - D_x^3 \cdot 135000 + 0.5.D_x^3 - 0.5.D_x^3 - D_x^2 = 0$$

$$135000D_x^3 + D_x^2 - 135000D_x = 0 \text{ solve}, D_x \rightarrow \begin{pmatrix} 0 \\ \dfrac{\sqrt{72900000001}}{270000} - \dfrac{1}{270000} \\ -\dfrac{\sqrt{72900000001}}{270000} - \dfrac{1}{270000} \end{pmatrix} = \begin{pmatrix} 0 \\ 1 \\ -1 \end{pmatrix}$$

Hence the price to maximise returns in a monopoly arises when demand is at a minimum. If the supply is continuous that is when supply approaches zero or, when demand is incremental, when the optimum supply is equal to 1. A demand of − 1 is outside the domain of demand and must therefore be ignored.

However even a product that in principle can be supplied in any quantity, such as a liquid, must by law be supplied in incremental quantities because of statutory units of measure. In practice therefore all products must be supplied incrementally and the optimum level of supply is 1 statutory unit of measure where:

$$P_x = V_x$$

While this level of supply will maximise profit, if our assumed demand function still applies to such a low volume, though this in principle maximises the monopolists return, it is not a very practical approach. Even a new business producing a product under monopoly conditions, due to patent protection, would be well advised to produce a much larger quantity, which provides a defensible market share when patent protection runs out.

Monopolists therefore, like oligopolists, have a strategic choice to make between a high price at a low volume or a lower price at a higher volume. The former will yield a higher return, but only limited market power and a high risk of competitive entry. The latter will provide a more dominant market position, and limit competitive entry but at the cost of lower returns. This choice is also represented graphically in figure 13 since the region of demand bounded by the demand function, y axis and the vertical black line at the competitive equilibrium, effectively represents the extent of the monopoly firm's freedom of manoeuvre if it is to do no worse than a perfectly competitive firm. We have however already demonstrated in the last chapter a form of optimum pricing for a monopoly in figure 11 that minimises unit cost. This however neither maximises returns nor minimises competitive entry but is a practical approach providing a defensible market share.

We might expect that in a relatively small private company, in which ownership is aligned with control and available capital is limited, holding a monopoly by patent protection for example, the firm would choose a higher price to maximise the principal's returns and minimise capital requirements. A large public company with diversified equity and much greater capital reserves, controlled by a professional management, may prefer to 'maximise' output and hence the power, prestige and income of the company and its management, thus preferring a lower price and higher volume.

Examples of both of these strategies for branded goods are to be found in the motor industry where a small niche market for very expensive prestige cars co-exists with a mass market for less expensive cars. The higher price of prestige cars partly reflects exclusivity and much higher quality, but an important reason why significantly higher prices can be maintained is the nature of monopoly pricing.

There is however a problem with the result of equation 27 in that the lower price could at best be below average cost as in our example. It is still possible for a firm to make a profit even when selling at below cost if they have a U shaped 'average' cost curve. The short term 'average' cost curve for most firms will tend to be U shaped. A profit can still be earned if the area below the equilibrium price line and above the direct cost curve is greater than the area/s below the direct cost curve and above the price line.

Some mass production or marketing firms do seem to operate on very low margins and may operate near such a low level of price particularly if market conditions are unfavourable. There is however no obvious reason why firms with monopoly power in more favourable circumstances would in general choose such a low price. We will return to resolve the issue of the rational choice of output and price for a large public company with monopoly power once we have considered the conditions for oligopoly pricing.

There is however an additional analysis we can undertake, which enables us to determine the level of output at which profitability per unit is greatest. Though additional gains can still be made by reducing output still further the profitability per unit will be lower at both a higher or lower level of output. Effectively reducing output further will yield diminishing returns.   This can be deduced from the second derivative of Equation 29.

Let:

$$f_6(D_x) := \frac{\left(f_1(D_x) - f_2(D_x)\right) \cdot 100}{f_1(D_x)}$$

and

$$f_7(D_x) := \frac{\left(f_1(D_x) - f_2(D_x)\right) \cdot 100}{f_2(D_x)}$$

If we plot $f_6(D_x)$ and $f_7(D_x)$ we obtain a graph of the following form:

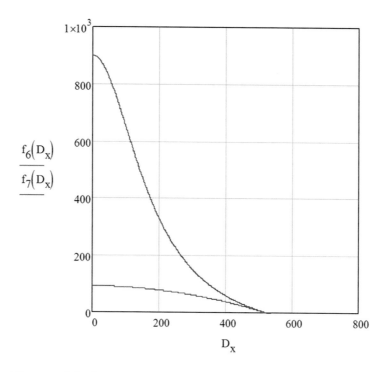

Figure.14 Graph Showing Profit as a % of Revenue in Red and Cost in Blue

The slope of $f_7(D_x)$ is never steeper (more negative) than when:

$$\frac{d^2}{d D_x{}^2} \left(f_7(D_x)\right) = 0 \text{ solve}, D_x \rightarrow \begin{pmatrix} -100.99833331944421296 \\ 98.99833331944421295 \end{pmatrix}$$

The 'ideal' range of output for a monopolist taking the data from our example is therefore between 98.998 to 3 decimal places, where profitability per unit is greatest and 173.205 to 3 decimal places, where cost per unit is least, giving a range of prices from £1,515.00 to £866.03 per unit respectively.

Both the two profitability functions intersect the x axis when demand is equal to supply in a competitive market. In the present context this occurs when:

$$\frac{\left(f_1(D_x) - f_2(D_x)\right) \cdot 100}{f_2(D_x)} = 0 \text{ solve}, D_x \rightarrow \begin{pmatrix} -520.6162045202208664 \\ 518.6162045202208664 \end{pmatrix}$$

$$\frac{\left(f_1(D_x) - f_2(D_x)\right) \cdot 100}{f_1(D_x)} = 0 \text{ solve}, D_x \rightarrow \begin{pmatrix} 0 \\ 518.6162045202208664 \\ -520.6162045202208664 \end{pmatrix}$$

The insight provided by the nature of monopoly pricing allows us to draw some more general conclusions about a firm's costs and pricing and not just under monopoly conditions either. In the short-run there is no such thing as normal profit. Profit is simply a residual from price after deducting costs. The price equation provides more of a challenge to a firm to get their level of costs below the market price rather than an expectation of profits. Even monopolists will be forced to accept prevailing market prices if they over estimate the size of the market and over produce or if the market moves against them. This is often figuratively described as selling at 'marginal' cost. That is at prices that are $< C_D$ AND $> C_V$ where a firm is otherwise unable to sell their output. If they still cannot shift their surplus stock selling at $< C_V$ will help to maintain cash flow.

This also explains the vulnerability of newly formed firms. A new firm will tend to have low levels of capitalisation. Thus, they are likely to have a very shallow U, shaped cost curve. Such firms therefore have very little tolerance of adverse market conditions. A very slight drop in the prevailing market price will make them unprofitable. In the short-term making a loss is the natural condition of a firm. Only those firms that have built up sufficient reserves and are effective in control of costs can weather unfavourable market conditions for very long.

There is however a major problem with this analysis based on an evaluation of the maximum monopoly profit because there is no optimum level of monopoly output within the domain of demand. The practical level of output to maximise profit is 1. This makes the analysis not only complicated but also under some circumstances unreliable. Despite this I will persist with this analysis because of its applicability in other forms of markets with different demand functions.

If a monopoly arises for a new product with patent protection for example this analysis need not be unreliable because such a new product could be manufactured at low volumes initially until the possible size of the market becomes clear. If however a monopoly is established for an existing product by one or more takeover bids, the prevailing level of output could be substantially reduced by the new owner to capitalise on their newly acquired monopoly powers. Under these circumstances the analysis is likely to become unreliable at very low levels of output.

Because we are basing our analysis on short term conditions when the model is applied to very low levels of output for a product previously sold in volume, these conditions become unsuitable. This is because in the short term though demand in a competitive market may be sensitive to price, the total number of consumers of the product is generally not likely to be significantly affected by a marginal reduction of the price. As preference and priority are learned behaviour, a 'marginal' increase in price just affects the quantity each consumer buys, not the number of consumers of the product, which will substantially remain unchanged.

In a competitive market such consumer's change of behaviour can only occur in the long or medium term because giving up a product altogether is learned behaviour. Also, as we will see later a sudden substantial rise in price can only occur in a speculative market not in a competitive market. Only when a substantial rise in price of a widely consumed good occurs will consumers with modest incomes need to quickly reassess their behaviour and give up consumption of a good altogether. This is the condition required to change a mass market good into a luxury good causing the number of consumers of a good to fall sharply. When both the number of consumers buying a good fall, as well as the amount each of the remaining consumers buys, the form of the demand function is likely to change. The market has in fact suddenly changed into a speculative market and our assumption that the demand function will not change, only the volume consumed, is no longer valid. This is something like the opposite condition to that described by (Thorstein Veblen, 1899).

Returning more specifically to monopoly conditions, it is generally held that monopoly involves a loss of consumer welfare due to constriction of output by the monopolist. For competitive goods as defined here this is clearly not so. The value of output under the demand curve of a competitive good is always P_xD_x which at any given time, with any given state of consumer preference and priority, will always be equal to $a_xYe^{-Y/N\lambda} + b_x$ or V_x. The value of competitive goods produced for a market (and the utility derived from their consumption) in a given state of preference is therefore always the same, irrespective of whether a small volume is produced at a high price or a large volume at a low price. Monopoly pricing therefore involves no direct loss of welfare to society as a whole, though a higher price effectively redistributes utility to the wealthy.

One reason often given for loss of welfare under monopoly is the reduction of consumer surplus. Under the new theory of value the concept of consumer surplus itself has no meaning since consumers always value their acquisitions at the opportunity cost of obtaining or retaining them. No consumer would ever be prepared to pay a higher price.

Consumer surplus (we will retain the name simply as a label), represented in figure 16 by the area beneath the demand function and above the horizontal blue line (prevailing market price), does however have a rather strange but different meaning under the new theory of value. It provides a direct way of evaluating the greatest happiness of the greatest number, allowing the utilitarian theories of (Jeremy Bentham,1789) to be largely rehabilitated.

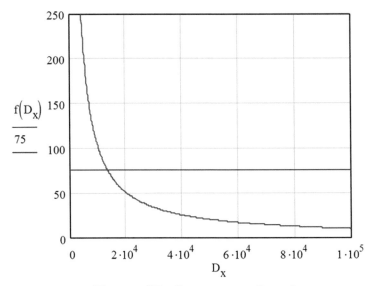

Figure 15. Consumer Surplus

The 'consumer surplus' arising from consumption of the first unit of production of any competitive good is indeterminate but that arising from all subsequent consumption can be evaluated as shown below. Incidentally the same argument applies to the evaluation of monopoly prices because the area under the entire demand function of a competitive good must always be indeterminate since the competitive demand function is an inverse (or reciprocal) function.

Let the 'value' of 'consumer surplus' arising from consumption of good x be V_{SX}, then:

$$V_{SX} = \int_1^{D_x} P_x dD_x - V_x$$

therefore

$$V_{SX} = \int_1^{D_x} \frac{V_x}{D_x} dD_x - V_x$$

Now at any given time with any given state of preference and priority, V_x is constant. Therefore:

$$V_{SX} = V_x \log_e D_x - V_x \qquad \text{Equation 30}$$

Where V_{SX} is the 'value' of 'consumer surplus' derived from consumption of good x.

Since at any point in time all consumers allocate a fixed proportion of their resources to a particular good, increasing the quantity of a good supplied to the market will not directly contribute to an increase in the number of consumers that will benefit from its consumption. However by the imitative behaviour of consumers, described by (Thorstein Veblen, 1899) the availability of a good, previously regarded as a luxury, in greater volumes at lower prices, will over time promote the adoption of that good into the consumption set of a wider group of consumers. Eventually therefore a greater value of V_{SX} promotes consumption by a greater number of consumers fulfilling the greatest happiness principle.

The problem with Bentham's formulation was that two variables cannot be optimised simultaneously in mathematics (except by chance). Hitherto therefore, we could not necessarily maximise utility and the number of consumers of a good simultaneously. However since the utility or value of goods consumed under any particular competitive demand function must always be the same in the short term the problem is reduced to one of maximising effective demand only. Since consumer surplus is proportional to demand and therefore to the number N_{cx} of consumers of good x it must also be proportional to 'happiness' or well-being, in the sense that the term is used by Bentham and incidentally in the case of the latter description by economists generally. This follows since $N_{cx} = D_x/q_{AVx}$, where q_{AVx} is the average consumption of good x. Furthermore, since $U_x = V_x$ consumer surplus is also proportional to utility. We will consider this argument further when we investigate welfare economics.

Examples of such goods that are now widely consumed in volume that were once the sole preserve of the rich are: home ownership, cars, consumer durables and foreign holidays. Most people would agree that, even though there is a downside to the emergence of mass markets for such goods, their availability has increased not reduced the sum total of human happiness.

Monopoly conditions therefore, though they do not involve a direct loss of welfare, work contrary to the greatest happiness principle and therefore monopoly remains undesirable unless unavoidable.

Domestic Borrowing

Monopoly pricing is not the only form of pricing solution involving the square root of a group of parameters generally associated with a 'quadratic' equation or those as in the case of monopoly conditions of higher order. All borrowing follows a similar rule. Domestic borrowing can be treated as just another consumption good. An individual's demand for borrowed money q_B like everything else is defined by the demand function for borrowing, which in conventional form is given by:

$$P_B = (\alpha_B y e^{-y/\lambda} + \beta_B)/q_B$$

Where
P_B is the price of borrowing
q_B is the amount borrowed
α_B is a consumer's preference for borrowing
and
β_B is a consumer's priority for borrowing

The price of money could be held to be the rate of interest and this would be the cost of a permanent overdraft. For other forms of borrowing consumers are more likely to regard the level of repayment of the loan as the 'price' of borrowing since this is the amount that must be allocated from income each month.

Repayment of a loan subject to compound interest on the outstanding balance is, from first principles, given by the formula:

$$R_L = \frac{q_B \cdot (1 + r)^n}{\sum_{m=0}^{n-1} (1 + r)^m}$$

where

R_L is the repayment of the loan per period
q_B is the principle amount borrowed
r is the rate of compound interest per period
n is the number of repayments

which simplifies to

$$R_L = q_B \cdot \frac{r \cdot (1 + r)^n}{\left[(1 + r)^n - 1\right]}$$

Equation 31

Note that this result is effectively a demand function for borrowing.

But $\qquad P_B \approx 12.R_L \qquad$ if repayments are made monthly

and $\qquad \beta_B = 0$

since it is nonsensical to finance borrowing from borrowing

therefore

$$q_B \cdot \frac{12r \cdot (1 + r)^n}{\left[(1 + r)^n - 1\right]} = \frac{\alpha_B \cdot y \cdot e^{\frac{-y}{\lambda}}}{q_B}$$

and

$$q_B = \sqrt{\alpha_B \cdot y \cdot e^{\frac{-y}{\lambda}} \cdot \frac{\left[(1 + r)^n - 1\right]}{12r \cdot (1 + r)^n}}$$

Equation 32

Note this result has both a positive and a negative root so equation 32 is a solution for lending as well as borrowing if lenders and borrowers have the same preference for contracting the loan. Should they not agree then no contract can be struck. If a consumer has multiple loans at different interest rates or over different terms all of these can be added together summing to β, his total indebtedness. Each term on the right-hand side of the equation will be of the same form as the right-hand side of equation 32. If a standing overdraft is also held, the sum between m = 0 and m = n − 1 (the term in r and n) reduces to approximately r/12 for this form of borrowing.

For a loan of a particular type x, defined as a loan at a particular rate of interest r over a particular repayment term n, repayments R_{Lx} are then as given below.

$$R_{Lx} = \sqrt{\alpha_{Bx} \cdot y \cdot e^{\frac{-y}{\lambda}} \cdot \frac{r \cdot (1+r)^n}{12 \cdot \left[(1+r)^n - 1\right]}} \qquad \text{Equation 33}$$

Consumers will generally prefer borrowing to all other goods. What limits their demand for borrowing is their ability to service the debt. Typically, their preference for almost any other good x will be $0 < \alpha_x < 1$ and generally very much closer to zero than one. Preference for borrowing, on the other hand is likely to be $\alpha_B \gg 1$ as we will now demonstrate. If a consumer has a tolerance for servicing a mortgage up to one third of his income we can deduce his preference for borrowing by solving equation 33 for α_B when:

$$R_{Lx} = y/36$$

Such a consumer with an income of £100,000 seeking a 30-year mortgage, at a rate of interest of approximately 5% per annum (actually 0.41% per month) will be prepared to support repayments of £2,778 per month. This gives a value for α_B of 286,985. The consumer's demand for borrowing under these circumstances is then £522,195. This result will provide a better test for lenders in deciding to whom to lend and how much to lend than those customarily used.

Lenders may however have a different (more conservative) tolerance for repayments as a proportion of income since if mortgage interest rates were to rise to over 15% as they did in the mid-1980s the same consumer with the same income and tolerance for debt could only support a loan of £223,306. This result gives an example of the borrower's and the lender's exposure to risk. Note also that the prevailing levels of interest rates early in the twenty first century have promoted a more intensely speculative market in house purchase than hitherto with much higher house prices. Should a significant rise in interest rates ever be necessary again the likelihood of very substantial negative equity for large numbers of house buyers and substantial bad debts would be very high, as was the case with sub-prime mortgages in the USA in the early 21st century.

From equation 32 we can deduce aggregate domestic borrowings D_{Bx} for each type of loan x across the whole economy as:

$$D_{Bx} = \sqrt{a_{Bx} \cdot Y \cdot e^{\dfrac{-Y}{N \cdot \lambda}} \cdot \dfrac{\left[(1+r)^n - 1\right]}{12r \cdot (1+r)^n}} \qquad \text{Equation 34}$$

Summing this result for all types of domestic borrowing provides a more predictable way of regulating domestic money supply or, more precisely, the demand for domestic borrowing, than that provided by the Quantity Theory of Money. This result is effectively a consumption function for borrowing of type x. A typical graph of this function, assuming aggregate preference for borrowing is the same as that assumed for an individual in the first example given above, is shown below.

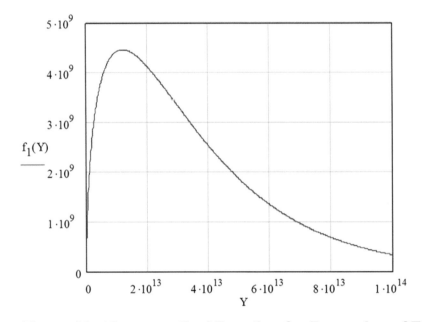

Figure 16. 'Consumption' Function for Borrowing of Type x

A similar analysis could be applied to commercial and government borrowing if we treat such organisations as if they were consumers.

Chapter 7. Oligopoly Markets

It is not surprising that the first theoretical framework in which bi-stable pricing was recognised was the new institutionalist analysis of oligopoly. This approach has one foot firmly grounded in a realistic description of oligopoly market conditions and the other in a formal analysis of the behaviour of the oligopoly firm. Armed with a 'neoclassical' analysis of apparently 'bi-stable' monopoly pricing we can attempt a synthesis of the two systems.

If we assume the collusive solution for oligopoly pricing achieves the lower level of demand and higher practical level of monopoly pricing, and the non-collusive solution achieves only the higher level of demand and lower practical level of monopoly pricing, we can represent all conditions of pricing in an oligopoly markets in a strategic form game as shown below.

It is assumed there are n firms in the market in two groups 0 to r and r+1 to n, where r can take any value from 0 to n - 1. Each firm chooses either a low volume strategy producing a market output of D_1 or a high-volume strategy producing a market output of D_4. Each firm then produces its market share δ_x of its chosen target market volume where x is an arbitrary firm in the range 0 to r or r+1 to n. The resulting market volume matrix for all firms in the market is then as given in table 1.

Firms 0 to r ⟍ Firms r+1 to n	$\delta_x D_1$	$\delta_x D_4$
$\delta_x D_1$	$D_1 \sum \delta_{\text{0 to r}}$ $D_1 \sum \delta_{\text{r+1 to n}}$	$D_1 \sum \delta_{\text{0 to r}}$ $D_4 \sum \delta_{\text{r+1 to n}}$
$\delta_x D_4$	$D_4 \sum \delta_{\text{0 to r}}$ $D_1 \sum \delta_{\text{r+1 to n}}$	$D_4 \sum \delta_{\text{0 to r}}$ $D_4 \sum \delta_{\text{r+1 to n}}$

Table 1. General Market Volume Matrix Under Oligopoly

There are now 4 different levels of output in each of the four cells of the matrix:

where D_1 is evaluated from $\dfrac{d^2 f_7(D_x)}{dD_x^2} = 0$ as above

$$D_2 = D_1 \sum_{0}^{r} \delta + D_4 \sum_{r+1}^{n} \delta$$

$$D_3 = D_4 \sum_{0}^{r} \delta + D_1 \sum_{r+1}^{n} \delta$$

and D$_4$ is evaluated from $\quad \dfrac{dC_D}{dD_x} = 0 \quad$ as above

The price matrix table is therefore as follows:

Firms 0 to r / Firms r+1 to n	$\delta_x D_1$	$\delta_x D_4$
$\delta_x D_1$	P$_1$	P$_2$
$\delta_x D_4$	P$_3$	P$_4$

Table 2. General Price Matrix Under Oligopoly

where

$$P_1 = \frac{V_x}{D_1}$$

$$P_2 = \frac{V_x}{D_2}$$

$$P_3 = \frac{V_x}{D_3}$$

$$P_4 = \frac{V_x}{D_4}$$

These levels of price and output can be represented schematically as shown in figure 17.

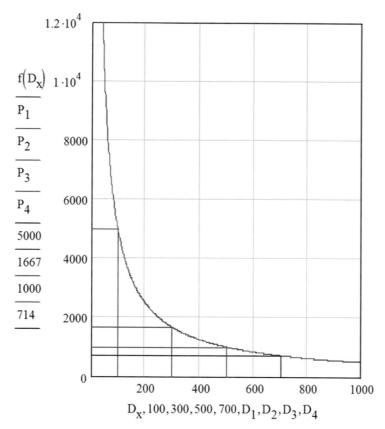

Figure 17. The Four Conditions of Oligopoly Pricing

Note that this graph has been constructed assuming equal bands of demand not equal bands of price, and the choice whether to represent $P_2 \geq$ or $\leq P_3$ is arbitrary depending on the values of $\sum\delta$ 0 to r and $\sum\delta$ r+1 to n since:

$$P_2 \geq \text{ or } \leq P_3 \quad \text{where 'or' has the common}$$

not the Boolean meaning

as
$$D_1\sum_0^r\delta + D_4\sum_{r+1}^n\delta \leq \text{ or } \geq D_4\sum_0^r\delta + D_1\sum_{r+1}^n\delta$$

now
$$\sum_0^r\delta = 1 - \sum_{r+1}^n\delta$$

and
$$\sum_{r+1}^n\delta = 1 - \sum_0^r\delta$$

therefore
$$D_1\sum_0^r\delta + D_4(1 - \sum_0^r\delta) \leq \text{ or } \geq D_4(1 - \sum_{r+1}^n\delta) + D_1\sum_{r+1}^n\delta$$

$$(D_1 - D_4)\sum_0^r\delta + D_4 \leq \text{ or } \geq (D_1 - D_4)\sum_{r+1}^n\delta + D_4$$

thus
$$P_2 \geq \text{ or } \leq P_3$$

as
$$\sum_0^r\delta \leq \text{ or } \geq \sum_{r+1}^n\delta \qquad \text{Relations 1}$$

The revenue matrix for each firm is shown in Table 3:

Firms 0 to r / Firms r+1 to n	δ_xD_1	δ_xD_4
δ_xD_1	$\delta_xD_1P_1$	$\delta_xD_4P_2$ $\delta_xD_1P_2$
δ_xD_4	$\delta_xD_4P_3$ $\delta_xD_1P_3$	$\delta_xD_4P_4$

Table 3. General Firms' Revenue Matrix Under Oligopoly

and the profits matrix for each firm is shown in Table 4.

Firms 0 to r / Firms r+1 to n	δ_xD_1	δ_xD_4
δ_xD_1	Z_1	$\angle_3$ Z_2
δ_xD_4	Z_4 Z_6	Z_6

Table 4. General Firms' Profit Matrix Under Oligopoly

Where
$$Z_1 = \delta_xD_1P_1 - C_D\delta_xD_1$$
$$Z_2 = \delta_xD_1P_2 - C_D\,\delta_xD_1$$
$$Z_3 = \delta_xD_4P_2 - C_D\,\delta_xD_4$$
$$Z_4 = \delta_xD_4P_3 - C_D\,\delta_xD_4$$
$$Z_5 = \delta_xD_1P_3 - C_D\,\delta_xD_1$$
and
$$Z_6 = \delta_xD_4P_4 - C_D\,\delta_xD_4$$

Note that none of these profit equations presents an evaluative difficulty for a firm if it can sell all its output since:

Where C_D is each firm's direct cost

and $\delta_x D_x$ is each firm's volume of sales for product x

Now we have a complete general description of how an oligopoly market works we can begin to examine its consequences. The general form of the matrices is asymmetric though all four matrices will be symmetrical if:

$$\sum_{0}^{r}\delta = \sum_{r+1}^{n}\delta$$

The system of four matrices, therefore describes not just all conditions in oligopoly markets but all conditions in monopoly and oligopoly markets. It is therefore a general solution for pricing in all non-competitive markets. The system of matrices reflects four levels of market concentration.

The top left hand quadrant represents pure monopoly, monopolistic competition between sharply differentiated branded goods or collusive oligopoly. This approach is appropriate for a small recently started monopoly firm enjoying monopoly by means of patent protection that seeks to maximise returns in the short term whilst establishing a defensible market share when patent protection runs out. The number of firms in the market n must therefore be very small. Collusion may be either implicit or overt.

The bottom right hand quadrant represents a non-collusive oligopoly market or a monopoly market dominate by a single large monopoly firm that seeks to minimise competitive entry into the market while achieving significantly greater return than can be achieved in a competitive market. The dominant firm may also wish to minimise the risk of government intervention by making modest profits compared with those gained by an exploitative monopoly. Where the market is a non-collusive oligopoly the number of firms in the market must therefore be large but not sufficiently large to promote perfect competition. As described in the last chapter, rather counter intuitively, the dominant firm may opt for the lower practical level of price available in a non-collusive oligopoly, while the weaker new firm, though they may wish to minimise costs initially, will, once firmly established, opt for the higher practical level of price achievable in a collusive oligopoly. One might have expected the opposite pairing. The reason is the strong monopoly firms seek to ward off the attentions of the regulator whereas the weaker new firm seeks to maximise its returns to build up its reserves before facing competitors when patent protection runs out.

When the number of firms in the market is small, but not sufficiently small to allow collusion, the matrices are asymmetric because as we shall see it is a matter of chance whether

$$\sum_{0}^{r}\delta \geq or \leq \sum_{r+1}^{n}\delta$$

If r firms adopt a niche market, low volume high price strategy and n − r firms adopt a mass-market, high-volume low-price strategy, then the pricing solution is given in the bottom left hand quadrant. If r firms adopt a mass-market strategy and n − r firms adopt a niche market strategy, the pricing solution is given in the top right-hand quadrant.

If n is large, but not so large as to promote perfectly competitive conditions, the matrix will tend to a symmetrical form. Since both strategies are open to all firms, markets with a large number of firms tend not to be completely dominated by any single firm and, as we shall see, either strategy could prove equally profitable.

The strategy each firm adopts can therefore be viewed as a matter of chance and thus, since n is large on the average:

$$\sum_{0}^{r} \delta \approx \sum_{r+1}^{n} \delta$$

The asymmetric matrices then approximates the following symmetrical form:

Firms 0 to r / Firms r+1 to n	$\delta_x D_1$	$\delta_x D_4$
$\delta_x D_1$	D_1	$\dfrac{D_1 + D_4}{2}$
$\delta_x D_4$	$\dfrac{D_1 + D_4}{2}$	D_4

Table 5. Market Volume Matrix When n is Large

Firms 0 to r / Firms r+1 to n	$\delta_x D_1$	$\delta_x D_4$
$\delta_x D_1$	P_1	P_2
$\delta_x D_4$	P_2	P_4

Table 6. Price Matrix When n Is Large

Where $\qquad\qquad\qquad\qquad P_2 = \dfrac{2V_x}{D_1 + D_4}$

Firms 0 to r / Firms r+1 to n	$\delta_x D_1$	$\delta_x D_4$
$\delta_x D_1$	$\delta_x D_1 P_1$	$\delta_x D_1 P_2$ $\delta_x D_4 P_2$
$\delta_x D_4$	$\delta_x D_1 P_2$ $\delta_x D_4 P_2$	$\delta_x D_4 P_4$

Table 7. Firm's Revenue Matrix When n is Large

Firms 0 to r / Firms r+1 to n	$\delta_x D_1$	$\delta_x D_4$
$\delta_x D_1$	Z_1	Z_2 Z_3
$\delta_x D_4$	Z_2 Z_3	Z_4

Table 8. Firm's Profit Matrix When n is Large

Where

$$Z_1 = \delta_x D_1 P_1 - C_D \delta_x D_1$$

$$Z_2 = \delta_x D_1 P_2 - C_D \delta_x D_1$$

$$Z_3 = \delta_x D_4 P_2 - C_D \delta_x D_4$$

$$Z_4 = \delta_x D_4 P_4 - C_D \delta_x D_4$$

All cells of the matrices in both the asymmetric and the symmetrical forms yield elevated prices and profits above the perfectly competitive level of price. These are generally referred to as super normal profits though the current theory, because it concentrates on short-term outcomes, does not recognise a normal level of profit.

We can now return to the unresolved problem of the level of output and price that a large publicly owned and professionally managed 'monopoly' firm will choose. Large firms are necessarily mass producers. The niche market solution given in the top left-hand quadrant is inappropriate for such a firm. A market dominated by a single firm will necessarily imply an asymmetric solution.

The rational choice for a large public company is therefore to produce an output of $\delta_x D_4$. This will require prices to be set at P_2 or P_3 whichever is the greater. This is the economic optimum, but see the argument above about choosing the uneconomic alternative to avoid the attentions of the regulator. Indeed a dominant firm may well choose to offer products at both levels of price since such firms often seek to stratify the market with different offerings pitched at different income groups. The most obvious example of this approach is once again to be found in the motor industry in which each firm offers multiple specifications in each product range. The policy is however typical of durable mass consumption products in general.

We could use this system to derive a similar set of matrices to define the conditions in all non-collusive oligopoly markets and all competitive markets from slightly competitive to perfectly competitive. Similarly, we could also develop a further set of matrices covering all monopoly conditions between the absolute optimum condition of supply for the monopolist and the sub-optimal condition to maximise profitability. The three sets of matrices would then cover all conditions in all markets. We could therefore using all three sets of matrices apply this approach to all levels of industrial concentration and all market conditions from exploitative monopoly to a perfectly competitive market or a real market approaching perfect competition.

Hence we could use this system to define an economic equilibrium, using any set or all sets of matrices, (that would cause markets to clear) but not necessarily (since I am not attempting to prove this here) an equilibrium in the sense used by (John Nash, 1950) (an outcome that would result if each firm pursued their own best interest irrespective of the decisions of others).

In part this arises from the property of monopoly markets that implies the optimum level of supply by a monopolist is 1 if the demand function of a competitive market still applies to such a low level of output. As a consequence, the monopolist or oligopolist has to compromise and choose one of two alternative sub optimal strategies. As a result it is unclear what a monopolist's or oligopolist's best interest is and hence Nash's test is more problematic.

Furthermore it follows that an oligopoly market may even be preferable to a perfectly competitive market from a public welfare perspective since firms pricing problem actually promotes a stratified market. This enhances consumer choice and hence consumer sovereignty. A perfectly competitive market on the other hand may actually promote adverse selection and a homogenous market.

Oligopoly markets are also advantageous from other perspectives. The size of firm under oligopoly allows greater advantage to be taken of economies of scale (compared with perfectly competitive firms) resulting in technical efficiency and lower costs. Oligopoly also avoids the fat, lazy firm syndrome that can arise under pure monopoly. Competition may even be fiercer when a firm knows its rivals. The perfectly competitive market's anonymous hordes of competing firms mean firms cannot know all their rivals.

Finally, competition between oligopoly firms promotes technical development and investment. Since most investment is funded from retained earnings and profits are normally higher in oligopoly firms than perfectly competitive firms, in the conventional sense, we might expect more investment in plant, training and R&D. Oligopoly therefore promotes growth and hence dynamic efficiency. We will consider how allocative efficiency is affected when we consider the general equilibrium of real markets.

All the matrices show higher oligopoly prices and profits arise in the top left-hand quadrant and lower oligopoly prices and profits in the top right or bottom left-hand quadrants. The collusive solution is however only available when the number of firms in the market is small and in any event a dominant firm, as we have seen, is unlikely to pursue a niche market strategy. For other firms, adopting either strategy does not necessarily imply higher or lower profitability since firms will seek to identify levels of activity and capitalisation appropriate to their chosen strategy. A monopoly or oligopoly firm, if well managed, may therefore be able to achieve higher profits following either strategy. It is this property of oligopoly markets that makes each firm's choice of strategy effectively a matter of chance.

Chapter 8. Speculative Markets

A speculation will develop when demand outstrips supply due either to restriction of supply or a rapid exogenous change of preference. Precisely why a speculative market arises however is not entirely clear, but it seems to result from the expectation of some consumers that a speculative market 'will' arise. If there are enough of such 'consumers' their expectation becomes a self-fulfilling prophecy.   These buyers ignore the market's signals to economise by reducing 'consumption' when prices rise. This behaviour could therefore be thought of as uneconomic if the object of speculation was viewed solely as a good. It is however precisely because the real value of an object of speculation rises over time, that the 'natural' functions of a good become confused with the wealth storing function of money that a speculation will arise. The function of money as a store of wealth then fails (at least with respect to the object of speculation) and at least initially the object of speculation becomes a better vehicle for the accumulation of wealth for those consumers with available liquid funds. Another way to express this is to say that some consumers have developed a time preference for the object of speculation over money.

When this occurs the value of λ in the consumption function for the object of speculation abruptly switches from positive to negative. In effect the object of speculation has taken on part of the wealth storing function of money. The 'consumption' function of the object of speculation then becomes an exponential growth function over time and with respect to income rather than a saturating function. Incidentally time represents an alternative calibration of the x-axis of the consumption function.

The form of the speculative 'consumption' function if the absolute value of λ remains unchanged is then as given below:

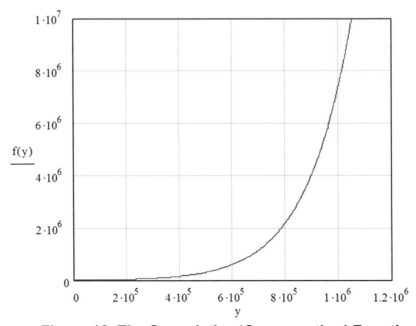

Figure 18. The Speculative 'Consumption' Function

In a speculative market consumers are forced to limit their search for the best value offerings even (indeed more so) for significant purchases, because to delay will involve a loss of disposable income or assets available for other purposes. Greater error then arises from an extended rather than a limited search. This effectively reduces the number of suppliers available to a consumer, allowing the suppliers to that market to charge monopoly prices even though the number of firms operating in the market may not normally allow monopoly pricing.

Speculative markets are virtual monopoly markets. The term virtual monopoly is sometimes used loosely to denote a market with a small number of firms. It is not used in that way here. A virtual monopoly market is defined as a market that would normally be a competitive market (in the conventional sense) but where suppliers are provided with an opportunity to charge monopoly prices because the market is a speculative market in which consumers must limit their search. The most obvious example of such a market in which many consumers participate is the house purchase market

The conditions for emergence of a speculative market may be derived from equation 20 as follows:

$$\frac{dP_x}{dD_x} = (D_x \frac{dV_x}{dD_x} - V_x)/D_x^2$$

$$\frac{dP_x}{dD_x} > 0 \quad \text{for an inverse demand function and therefore}$$

$$(D_x \frac{dV_x}{dD_x} - V_x)/D_x^2 > 0$$

$$D_x \frac{dV_x}{dD_x} > V_x$$

now $$\frac{dV_x}{dD_x} = \frac{dV_x}{dY} . \frac{dY}{dD_x} \quad \text{by the chain rule}$$

therefore $$D_x \frac{dV_x}{dY} > V_x \frac{dD_x}{dY}$$

thus for an individual to participate in a speculative market

$$q_x(\alpha_x e^{-y/\lambda} - \frac{\alpha_x y e^{-y/\lambda}}{\lambda}) > (\alpha_x y e^{-y/\lambda} + \beta_x)\frac{dq_x}{dy} \quad \text{Relation 2}$$

and $$q_x(\frac{1}{y} - \frac{1}{\lambda}) > (\frac{\beta_x}{\alpha_x y e^{-y/\lambda}} + 1)\frac{dq_x}{dy} \quad \text{Relation 3}$$

Note these relations, though based on the market demand function, are formulated with respect to individual income, not National Income. Why this was done will become apparent from the following. The conditions for an individual's participation in a speculative market are therefore all those listed below:

$$\frac{dq_x}{dy} > 0 \qquad \text{demand increases with income}$$

$$q_x \; > 0 \qquad \text{demand is positive}$$

$$\beta_x \; \to 0 \qquad \text{priority is low}$$

$$\alpha_x \; > 0 \qquad \text{preference is high}$$

$$e^{-y/\lambda} > 0 \qquad \text{always}$$

$$\left(\frac{\beta_x}{\alpha_x y e^{-y/\lambda}} + 1\right) > 0 \qquad$$ for good x to be consumed at all because the left-hand side of this expression can be reformulated as the consumption function for good x

therefore $$\left(\frac{1}{y} - \frac{1}{\lambda}\right) > 0$$

If an individual's income is less than the modulus of λ, this condition will hold irrespective of whether λ is positive or negative. If an individual's income is greater than the modulus of λ this condition only holds when λ is negative. There is no reason to suppose that high earners are less likely to speculate than low earners, on the contrary. they are more likely to, since they have higher disposable incomes and can better afford to take the risk. The latter condition must therefore apply.

thus $$\lambda < 0 \qquad\qquad \text{Condition 1}$$

This condition that $\lambda < 0$ is the fundamental characteristic of a speculative market. If it can ever be shown that λ is negative for a particular good it follows that that good must be traded in a speculative market.

Since virtual monopoly conditions apply, we may assume that the form of the speculative demand function can be derived from the condition for maximising monopoly profits, the competitive demand function and parameters that arise at equilibrium in a competitive market. The assumption of these parameters may be considered reasonable because monopoly prices depend on the assumption of a competitive demand function.

From equation 26: $$D_x \frac{dP_x}{dD_x} = C_D - P_x$$

then, to a first approximation, substituting for P_x from a form of equation 19:

$$\frac{D_x dP_x}{dD} = C_D - \frac{V_x}{D_x}$$

dividing by D_x

$$\frac{dP_x}{dD_x} = \frac{C_D}{D_x} - \frac{V_x}{D_x{}^2}$$

therefore

$$\int_{V_x}^{P_x} .dP_x = \int_{1}^{D_x} \frac{C_D}{D_x} dD_x - \int_{1}^{D_x} \frac{V_x}{D_x{}^2} dD_x$$

Where C_D is assumed to be the direct cost at the competitive equilibrium of good x a proxy variable for the competitive equilibrium price
P_x is the price in a speculative market at any level of demand ≥ 1
and V_x is the competitive level of consumption

Note the limits applied to integration in this equation are necessary because $\log_e D_x$ is indeterminate when D_x is zero. This does not pose any practical problem provided unit quantity of good x is defined as the minimum amount of x that can be purchased. The minimum quantity of any good that may be purchased is always either 1 or the minimum statutory unit of measure for that good. For example, petrol (gas) is sold at the pump to two decimal places of a litre in the UK. Unit quantity of petrol must therefore be defined as one centilitre if this equation were ever applied to a speculative market for petrol on filling station forecourts in the UK.

As V_x is the level of consumption that would arise in a competitive market the price of good x when demand is one is also V_x.

Then: $$P_x - V_x = C_D \log_e D_x + \frac{V_x}{D_x} - V_x$$

Thus, the terms in V_x only, cancel one another out and:

$$P_x = C_D \log_e D_x + \frac{V_x}{D_x} \qquad \qquad \text{Equation 35}$$

Using this analysis the emergence of a speculative market therefore increases prices by $C_D \log_e D_x$. The form of the speculative demand function based on this equation is shown below:

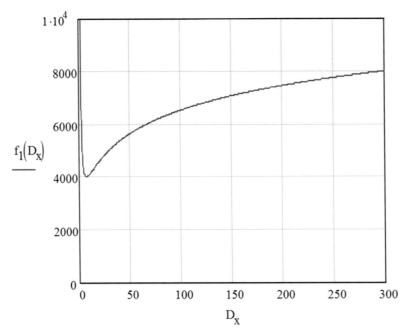

Figure 19. The Speculative Demand Function

The shape of figure 19, derived mathematically, may be considered to follow reasonable expectations of the form of the speculative demand function because at levels of demand below the competitive equilibrium it is similar to the competitive demand function and at higher levels of demand it is an inverse demand curve, that is a reasonable compromise between the competitive demand function of figure 7 and the speculative 'consumption' function of figure 18. If the argument about C_D being the direct cost at the competitive equilibrium (a proxy variable for the competitive equilibrium price) is thought to be unconvincing then this derivation creates a difficulty since it assumes C_D is constant.

The problem with any other assumption however is, that there is no generally agreed or universally applicable formula for marginal cost. We have therefore made what is perhaps a more realistic assumption (based on the accounting concept of direct cost) that short-term unit cost follows a function of the form given in Equation 24 as shown below:

$$C_D = \frac{C_F}{q_x} + C_V + C_E.q_x$$

Where:
    q_x is the quantity of good x produced
    C_F is fixed or capital cost
    C_V is variable cost

and
    C_E is exceptional costs such as overtime payments, shift premiums, short term purchasing premiums or discounts lost.

Figure 20 below shows the form of the direct cost function with the parametric conditions assumed above.

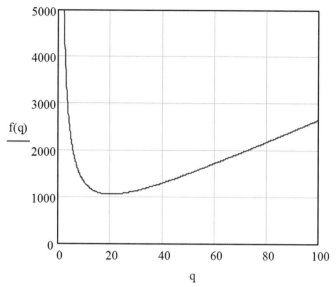

Figure 20. A Possible U Shaped Direct Cost Curve

From equation 24 we can calculate average cost by integrating C_D between 1 and D_x and dividing by D_x. The graph below shows the short run direct cost curve of figure 20 in blue, average cost in black and the associated competitive demand function in red. Note that average cost at equilibrium exceeds the competitive equilibrium price in this case.

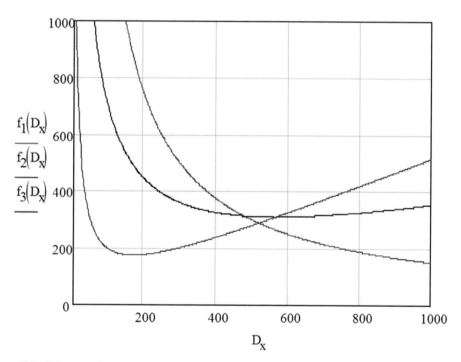

Figure 21. Direct Cost Average Cost & Competitive Demand Functions

If we assume equation 24 applies then substituting for C_D in the integral equation above we have:

$$D_x \frac{dP_x}{dD_x} = \frac{C_F}{D_x} + C_V + C_E . D_x - P_x$$

$$\int_{V_x}^{P_x} . dP = \int_{1}^{D_x} \frac{C_F}{D_x^2} + \frac{C_V}{D_x} + C_E - \frac{P_x}{D_x} \, dD_x$$

substituting for P_x
$$\int_{V_x}^{P_x} . dP = \int_{1}^{D_x} \frac{C_F}{D_x^2} + \frac{C_V}{D_x} + C_E - \frac{V_x}{D_x^2} \, dD_x$$

then
$$P_x - V_x = C_V \log_e D_x + \frac{(V_x - C_F)}{D_x} + C_E(D_x - 1) - (V_x - C_F)$$

$$P_x = C_V . \log_e D_x + \frac{(V_x - C_F)}{D_x} + C_E (D_x - 1) + C_F \quad \text{Equation 36}$$

If we add a true cost correction parameter t_c to the first term in equation, 35 and plot this modified form of equation 35 and equation 36 together for values of the parameters calculated according to the procedure defined in the following section we typically obtain the result shown below.

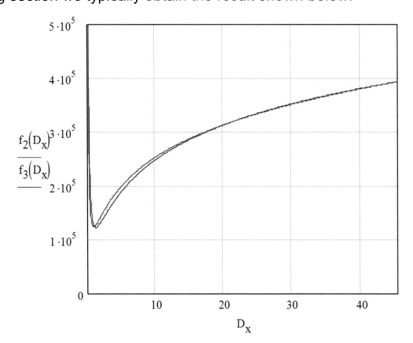

Figure 22. 'True' & Approximate Speculative Demand Functions

The 'true' speculative demand function plotted from equation 36 based on 'realistic' assumptions concerning the direct cost function is shown in red, and the corrected approximate speculative demand function plotted from equation

35 based on the assumptions described below is shown in blue. Both curves are plotted across the predicted domain of D_x also described below.

Under these assumptions the approximation achieves a reasonably close fit to the true speculative demand function without the need to evaluate the costs C_F, C_V or C_E, which would differ from firm to firm, from good to good and from time to time. Though the approximation may not always be as close as this example (the approximation is poor when $2V_x/C_D$ is small) it will generally be reasonable. Note that this form of speculative inverse demand function is continuous with an approximately competitive demand function at levels of demand below the competitive equilibrium price.

Reformulating equation 35 to incorporate the appropriate value of direct cost gives:

$$P_x = t_c.C_D\log_e D_x + \frac{V_x}{D_x} \qquad \text{Equation 37}$$

Where t_c is a true cost correction factor selected to provide the closest fit of equation 35 to equation 36

We will show how to calculate t_c after we have examined the domain of demand of a speculative good for reasons that will become apparent. Though the approximation derived above achieves a reasonably close fit over a wide range of values for D_x the limited domain of demand of a speculative market improves the effectiveness of the approximation.

We can calculate the upper bound of the domain of D_x approximately from equation 27.

Since
$$\frac{dP_x}{dD_x} > 0$$

and thus
$$4V_x\frac{dP_x}{dD_x} > 0$$

then if we assume this expression is unbounded the maximum value it may take before the roots of equation 27 would become complex is given when:

$$C_D{}^2 = 4V_x\frac{dP_x}{dD_x}$$

Substituting for $\frac{dP_x}{dD_x}$ from page 69 above:

$$C_D{}^2 = 4V_x\left(\frac{C_D}{D_x} - \frac{V_x}{D_x{}^2}\right)$$

therefore
$$C_D{}^2 D_x{}^2 - 4V_x C_D D_x + 4V_x{}^2 = 0$$

dividing by $C_D{}^2$
$$D_x{}^2 - \frac{4V_x D_x}{C_D} + \frac{4V_x{}^2}{C_D{}^2} = 0$$

factorising
$$\left(D_x - \frac{2V_x}{C_D}\right)^2 = 0$$

and therefore $\quad\quad\quad\quad\quad D_x = \dfrac{2V_x}{C_D} \quad\quad\quad\quad\quad$ Equation 38

The domain of demand of a speculative market is therefore estimated to be the interval 1 to $2V_x/C_D$. This result is potentially very important since it suggests we may be able to predict the catastrophic failure of a speculative market.

However, as we shall see, this is only an approximate predictor of the upper bound of demand in a speculative market. Though equation 27 (the monopoly price equation) and equation 35 (the speculative demand function) are both derived from equation 26 (the condition for monopoly profits), equation 26 is only an approximation to equation 35. Furthermore though equation 26 is a close approximation to equation 35 at low levels of demand it is a poor approximation near to the upper bound of the domain of demand of a speculative good. We will however show later (when considering the partial equilibrium of speculative goods) how to calculate the upper bound of the domain of demand of a speculative good exactly in any particular case and that demand and hence price can still be unreal in a speculative market.

For the time being we will assume equation 38 is a reasonable predictor of the upper bound of the domain of demand in a speculative market. This is not an unreasonable assumption. We will show later that the upper bound of the domain of demand of a speculative good typically, though not invariably, lies within the interval $(V_x/C_D, 3V_x/C_D)$.

The true cost correcting parameter t_c can be evaluated as follows. If we define $f_1(D_x)$ as equation 35 (the approximate speculative demand function assuming direct cost is constant), and $f_2(D_x)$ as equation 36 (the 'true' speculative demand function based upon realistic assumptions concerning direct cost) and $f_3(D_x)$ as equation 37 (the corrected approximate speculative demand function) and solve the equation:

$$f_1(D_x) = f_2(D_x)$$

for D_x we find that D_x must always be equal to 1 for this condition to hold irrespective of the value of the parameters. Substituting this value for D_x into this equation gives the appropriate value of C_D required in equation 37 as:

$$C_D = C_F + C_V + C_E - V_x$$

Note this correction will only work if:

$$C_F + C_V + C_E > V_x$$

that is for capital intensive production processes. Assuming this value of C_D if we now solve the equation:

$$f_2(D_x) = f_3(D_x)$$

for t_c when $D_x = 2.V_x/C_D$ (the estimated upper bound of the domain of demand of a speculative market) we obtain the result:

$$t_c = \frac{\left[C_V\left(2\frac{V_x}{C_D}\right)\cdot\ln\left(2\frac{V_x}{C_D}\right) - C_F + C_E\left(2\frac{V_x}{C_D}\right)^2 + C_F\left(2\frac{V_x}{C_D}\right) - C_E\left(2\frac{V_x}{C_D}\right)\right]}{2V_x\cdot\ln\left(2\frac{V_x}{C_D}\right)}$$

These are the conditions used to plot figure 23. The value of the parameter t_c can be significantly different from 1.

For simplicity hereafter we may assume that C_D takes the value:

$$C_D = t_c.(C_F + C_V + C_E - V_x)$$

The important point about this result is that there is a value of C_D that makes equation 37 a close approximation to equation 36. In principle the value of C_D satisfying this condition could be estimated econometrically from equation 36, though not easily, because this would require non-linear regression, and estimating marginal cost, the preferred cost parameter, is notoriously difficult, hence our use of direct cost as an alternative. Also, speculative markets are often subject to a great deal of noise. An alternative to the conventional econometric method of estimating C_D is derived below.

The value of C_D is assumed, henceforth, to have been corrected in accordance with the procedure described above. This enables us to use the original form of speculative demand function of equation 35 where necessary instead of that of equation 36 or 37, though under different assumptions, throughout the rest of this work.

Irrespective of the assumptions made, a speculative bull market will only ever emerge at levels of demand above V_x/C_D, since $P_x \approx C_D$ is assumed to be the equilibrium condition of a competitive market even in the short term. If P_x were greater than C_D a firm would simply sell more, and if P_x were less than C_D they would sell less. For prices to be elevated above the competitive equilibrium price requires that demand is elevated above the competitive equilibrium level of demand and thus above V_x/C_D. Based on these assumptions the domain of a speculative bull market can therefore be considered to be approximately $(V_x/C_D, 2V_x/C_D)$. Since only bull markets have inverse demand curves, the interval $(V_x/C_D, 2V_x/C_D)$ defines the approximate limiting conditions for a speculative inverse demand curve.

Estimating the Parameters V_x and C_D

Once a speculative market has developed it may be quite difficult to establish the competitive level of consumption required in equations 35 to 39. Furthermore, as we have already noted, marginal cost is notoriously difficult to estimate, so establishing an appropriate cost could be equally difficult. We can however estimate these parameters from conditions prevailing after a speculative market has emerged as follows.

If
$$V_S = P_S \cdot D_S$$

where V_S is the speculative level of 'consumption'
               P_S is the price in a speculative market
and D_S is the speculative level of demand

then
$$\frac{d}{dD_S} V_S = P_S + D_S \cdot \frac{d}{dD_S} P_S$$

Let
$$f(D_S) = \frac{d}{dD_S} P_S$$

but
$$V_S = C_D \cdot D_S \cdot \ln(D_S) + V_C$$

where V_C is the competitive level of consumption

therefore
$$\frac{d}{dD_S} V_S = C_D \cdot \ln(D_S) + C_D$$

Thus, using the MATHCAD solve block (a procedure for solving systems of simultaneous equations and constraints):

given
$$P_S + D_S \cdot f(D_S) = C_D \cdot \ln(D_S) + C_D$$

and
$$P_S = C_D \cdot \ln(D_S) + \frac{V_C}{D_S}$$

$$\text{Find}(V_C, C_D) \rightarrow \begin{pmatrix} \dfrac{D_S \cdot P_S - D_S^2 \cdot f(D_S) \cdot \ln(D_S)}{\ln(D_S) + 1} \\ \dfrac{P_S + D_S \cdot f(D_S)}{\ln(D_S) + 1} \end{pmatrix}$$

Therefore
$$V_C = -D_S \cdot \frac{\left(-P_S + \ln(D_S) \cdot D_S \cdot f(D_S)\right)}{\left(\ln(D_S) + 1\right)}$$

and
$$C_D = \frac{\left(P_S + D_S \cdot f(D_S)\right)}{\left(\ln(D_S) + 1\right)}$$

If price and demand in a speculative market are measured on two consecutive days (or weeks or months in a more gradually rising speculative market):

for
$$P_1 < P_2$$

then
$$P_S = \frac{P_1 + P_2}{2}$$

and for
$$D_1 < D_2 \qquad \text{in a speculative bull market}$$

then
$$D_S = \frac{D_1 + D_2}{2}$$

and
$$f(D_S) = \frac{P_2 - P_1}{D_2 - D_1}$$

Therefore:

$$V_C = \left(\frac{1}{2} \cdot D_1 - \frac{1}{2} \cdot D_2\right) \cdot \frac{\frac{-1}{2} \cdot P_2 - \frac{1}{2} \cdot P_1 + \ln\left(\frac{1}{2} \cdot D_1 + \frac{1}{2} \cdot D_2\right) \cdot \left(\frac{1}{2} \cdot D_1 + \frac{1}{2} \cdot D_2\right) \cdot \frac{P_2 - P_1}{D_2 - D_1}}{\left(\ln\left(\frac{1}{2} \cdot D_1 + \frac{1}{2} \cdot D_2\right) + 1\right)}$$

and

$$C_D = \frac{\left[\frac{1}{2} \cdot P_1 + \frac{1}{2} \cdot P_2 + \left(\frac{1}{2} \cdot D_1 + \frac{1}{2} \cdot D_2\right) \cdot \frac{P_2 - P_1}{D_2 - D_1}\right]}{\left(\ln\left(\frac{1}{2} \cdot D_1 + \frac{1}{2} \cdot D_2\right) + 1\right)}$$

This method of estimating the parameters of equation 35 is very accurate provided it is made soon after a speculative market first develops, that is near to the competitive equilibrium if we assume the conditions used to derive equation 35 originally. Furthermore, because marginal cost is so notoriously difficult to estimate from a firm's accounts, the resulting value of unit cost is likely to be far more accurate than a firm could estimate its own marginal cost. In any event the value of C_D used in these equations is actually an aggregate (or form of average) direct cost for the whole market. Note however these estimates of parameters can be very inaccurate when made later in the development of a speculative market if we maintain the original assumptions of equation 35.

However because such estimates are made from prevailing conditions in a speculative market, we may assume that the corrections used to derive equation 37 already apply. An estimate made later in the development of a speculative market may therefore provide a better estimate of C_D, if we assume its optimisation to achieve the best approximation of equation 36 to equation 37.

This method of estimating conditions for a particular type (or better still model) of house in a speculative housing market in a particular area is likely to work quite well because house prices tend to rise in a 'smooth' progressive manner. In stock or commodity markets however prices are very erratic. These markets are under damped and therefore oversensitive markets that suffer from a great deal of noise due to jittery changes of sentiment or preference. It is likely in such markets that a number of estimates will therefore need to be made over a short period and averaged to deduce likely values of V_C and C_D. It would also be difficult to establish when the market was near to competitive equilibrium conditions. The best way to judge this is to estimate the parameters when the market appears to have begun to settle into bull market conditions after a long period of bear market conditions. Note that the original nominal or offer price on flotation of shares would be no guide to the 'true cost' of stocks and shares (except for newly issued shares) because they may have passed through many hands subsequently. What we are trying to estimate is the aggregate 'unit cost' to current owners, not to the original owners. Note because stock and commodity markets are so unstable equation 35 can at best only ever predict the underlying trend line of prices in these markets, assuming preference has not changed, not the short term price. In other forms of speculative market the predictions of equation 35 are likely to be more reliable.

Since in a speculative market the consumption and hence the demand function is a growth function and therefore the market is subject to inverse demand and the slope of the demand curve

$$\frac{dP_x}{dD_x} > 0$$

This does however allow some further elaboration of the speculative demand function to be derived, having a more rigorous basis.

Since the aggregate consumption function for good x in a competitive market is given by Equation 6:

$$V_x = a_x Y e^{-Y/N\lambda} + b_x$$

and in a speculative market the aggregate consumption function takes the form

$$V_x = a_x Y e^{Y/N\lambda} + b_x$$

because λ is negative. The relationship between consumption in a speculative market to that of a competitive market is therefore given by:

$$\frac{\left(a_x \cdot Y \cdot e^{\frac{Y}{N \cdot \lambda}} + b_x\right)}{\left[a_x \cdot Y \cdot e^{-\left(\frac{Y}{N \cdot \lambda}\right)} + b_x\right]}$$

In a speculative market priority b_x must be very low tending to zero because it makes no sense to give priority to speculation over staple goods or other regular household purchases, particularly for people with modest or middle incomes, or to borrow to speculate. Furthermore, most regular household purchases are less likely to be purchased from borrowing by most people. The ratio between consumption in the two markets therefore reduces to:

$$e^{\frac{2.Y}{N\lambda}}$$

In most cases, but Y/N is average income y_{AV} therefore the ratio between consumption in a speculative market to consumption in a competitive market reduces still further to:

$$e^{\frac{2.y_{AV}}{\lambda}}$$

The provisional figure for the average income of 'All adults in full time employment, in April 2017' was reported by the (Office for National Statistics, 2017) in 'The Annual Abstract of Statistics 2017, Table 1.4' to be £662.50 per week. This equates to £34,582.50 per annum at 2017 prices, assuming a 52.2 week year. However, the increase in the value of λ based on the compounded RPI between 2000 and 2017 obtained from Table 1.2 of the 2017 Abstract is proportionately 1.62397. The value of λ in April 2017 is therefore £350,000 to 3 significant figures if calculated in the same way as on page 19 above. Thus, the proportionate increase in consumption between 2000 and 2017 is 1.198 to 3 decimal places. This would increase the value of consumption of good x in our example from £150,000 in 2000 to £179,642.14 or approximately 180,000 to 3 significant figures were a speculative market to develop for that good in April 2017.

Incorporating this result, we can represent the demand function in a speculative market as

$$P_x = C_D \log_e D_x + (e^{2y_{AV}/\lambda}) \frac{V_x}{D_x} \qquad \text{Equation 39}$$

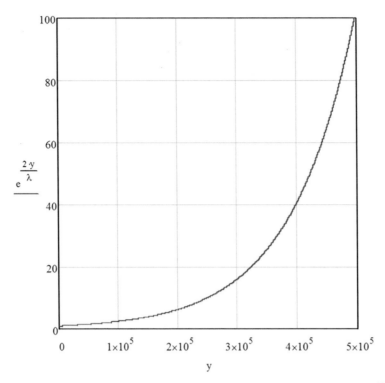

Figure 23. Graph showing how an individual's expenditure increases with income in a speculative market

In a speculative market we may therefore assume that if the speculation continues the terms under the square root sign of equation 27 would eventually become negative. Equation 27 would then have no real roots and there would therefore be no real prices in the market. If virtual monopoly conditions apply and there are multiple vendors, assuming no collusion, each supplier would then unilaterally set arbitrary prices resulting in multiple prices for essentially similar goods. The function of money as a measure of relative scarcity and value would then fail. Consumers would no longer be able to make rational 'utility maximising' choices and the speculative market would fail and collapse in chaos. The term failure is used here to mean a cataclysmic collapse of a market rather than, in the more usual sense in which this term is used in economics, as a failure of a market to promote public welfare.

It follows that there could be two ways in which a speculation can end. It could either peter out implying an exogenous change of preference or sentiment and a progressive turning point returning the market to competitive conditions or, it could end due to market failure as described above. The former will result in a terminating saturation phase in the consumption function at the end of the speculation. The latter implies a discontinuous turning point in the consumption function that is an endogenous, not an exogenous change. Consumers have not changed their preferences the market has just failed to supply them.

Though cases of the cataclysmic failure of speculative markets are spectacular and famous, they are fortunately rare. Examples are the 17th century Dutch bulb speculation, the South Sea bubble, the Wall Street crash, the 1992 collapse of the UK housing market, and the fall of technology shares after the millennium. This rarity is because buyers know that such a collapse can occur and confidence in the markets' continuing rise evaporates as prices rise ever higher. Sentiment or preference therefore tends to wane, producing a progressive rather than an abrupt turning point.

Losses can still be made but they are not as great as those arising from a cataclysmic failure, and neither do they affect as many people. Buyers are thus protected from the worst effects of cataclysmic failure by their own caution. Only in very rapidly rising markets, where the gains that can be made by those that entered the market early and sell before the fall are very great, does reckless speculation arise. The condition for a rapidly rising market ($\lambda_x << \lambda$) is illustrated in figure 26 below and discussed when we investigate the maximum level of useful income in a speculative market

If true monopoly conditions apply or if monopolistic competition applies, to sharply differentiated branded goods in a speculative market, the single supplier would be able to ensure unambiguous prices are maintained, even though they are arbitrary. The function of money, as a measure of relative scarcity and value, does not fail under these conditions and the absence of real prices in the market becomes sustainable.

Conspicuous Consumption

Conspicuous consumption is generally attributed to the desire of the rich to display their wealth (Thorstein Veblen, 1899). We all like to show off new acquisitions and the rich, particularly the nouveau riche, perhaps more motivated than most by pursuit of the material, could well gain greater utility from display. This however is not their primary motivation. Conspicuous consumption is a speculation in utility.

Some goods that are the object of conspicuous consumption may also gain real value over time providing a double benefit for the rich. Most conspicuous consumption however, involves goods like the most expensive brands of designer goods with a negligible resale value so the object of their purchase is not a growth of real assets. There must therefore be a different motive for conspicuous consumption.

The very rich face completely satiating consumption functions for all the goods in their consumption set and even, if they are rich enough, for their consumption overall. This is one reason why the rich, particularly the idle rich, often profess boredom with all consumption and a general disillusion with life.

The only goods that enable them to add to their utility significantly are the goods in speculative markets. Since the availability of unique goods like works of art or other rare items is low, the rich need a wider class of exclusive goods in which to 'speculate'. The most expensive brands of designer goods, like Versace dresses for example, satisfy this need. Since the residual value of designer garments and many other conspicuous consumption goods is negligible, the rich are directly speculating in their utility. The conditions for conspicuous consumption are derived from relations 2 and 3.

$$q_x(\alpha_x e^{-y/\lambda} - \frac{\alpha_x \underline{y} e^{-y/\lambda}}{\lambda}) > (\alpha_x y e^{-y/\lambda} + \beta_x)\frac{dq_x}{dy}$$

and

$$q_x(\frac{1}{y} - \frac{1}{\lambda}) > (\frac{\beta_x}{\alpha_x y e^{-y/\lambda}} + 1)\frac{dq_x}{dy}$$

For conspicuous consumption:

$\dfrac{dq_x}{dy} > 0$ demand increases with income

$q_x > 0$ demand is positive

$y > |\lambda|$ for genuinely conspicuous consumption as opposed to the imitative behaviour of less wealthy consumers

$\beta_x \rightarrow 0$ priority is low

$\alpha_x > 0$ preference is high

$e^{-y/\lambda} > 0$ always

$(\dfrac{\beta_x}{\alpha_x y e^{-y/\lambda}} + 1) > 0$ for good x is to be consumed at all.

therefore

$$(\frac{1}{y} - \frac{1}{\lambda}) > 0$$

and

$$\lambda < 0$$

This is the condition for a speculative market proving that conspicuous consumption is a speculation. Note this condition holds even if $\beta_x > 0$. Since the residual value of conspicuous consumption goods is not necessarily enhanced by the speculation but the opportunity cost of obtaining them (and therefore their utility) is, we must conclude conspicuous consumption is a direct speculation in utility.

The effect on a wealthy consumers' marginal utility of income due to the emergence of a speculative market for a conspicuous consumption good can be calculated as follows:

In a speculative market where $\lambda < 0$ $\dfrac{dU_{xs}}{dy} = a_x e^{y/|\lambda|} + \dfrac{a_x \underline{y} e^{y/|\lambda|}}{|\lambda|}$

In a competitive market where $\lambda > 0$ $\dfrac{dU_{xc}}{dy} = a_x e^{-y/|\lambda|} - \dfrac{a_x \underline{y} e^{-y/|\lambda|}}{|\lambda|}$

Subtracting marginal utility of income in a competitive market for good x from marginal utility of income in a speculative market for good x gives:

$$\Delta_{s-c}\frac{dU_x}{dy} = a_x(e^{y/|\lambda|} - e^{-y/|\lambda|}) + \frac{a_x y}{|\lambda|}(e^{y/|\lambda|} + e^{-y/|\lambda|})$$

$$\Delta_{s-c}\frac{dU_x}{dy} = 2a_x((\sinh(y/|\lambda|) + \frac{y}{|\lambda|}\cosh(y/|\lambda|)) \qquad \text{Equation 40}$$

Where $\Delta_{s-c}\frac{dU_x}{dy}$ is the increase in marginal utility of income arising from the consumption of good x due to the emergence of a speculative market, if we assume that the absolute value of λ does not change when a speculative market emerges. Plotting this equation for a good that represents initial consumption of 5% of income, that is where $a_x = 0.05$, results in a graph of the following form.

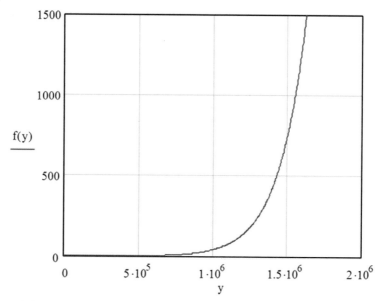

Figure 24. Increase in Utility Due to Conspicuous Consumption

This graph shows that the increase in marginal utility of income arising from consumption of good x is negligible at incomes below λ where utility increases by only 27p. At an income of £1.6M (approximately 8λ) however, utility would be increased by £1,341 if the domain of demand were unbounded. Thorstein Veblen was amazingly prescient when, without the benefit of this analysis, he 'differentiated' conspicuous consumption as applying only to the very rich and a distinct imitative form of consumption applying to the merely prosperous.

It is likely however that the absolute value of λ for a speculative good will be very much smaller than the value for a currency, particularly a hard currency. We can calculate the value of λ for a speculative good as follows:

$$C_x \cdot D_x \cdot \ln(D_x) + V_x = a_x \cdot Y \cdot e^{\frac{Y}{N \cdot \lambda_x}} + b_x \qquad \text{from equations 6 and 35}$$

$$\lambda_x = \frac{Y}{\ln\left[\dfrac{\left(C_x \cdot D_x \cdot \ln(D_x) + V_x - b_x\right)}{a_x \cdot Y}\right] \cdot N}$$

Where $\qquad b_x \to 0 \qquad\qquad$ since priority is low

A typical effect of a specific value of λ_x for a speculative good is shown in the graph below in which the absolute value of λ_x for a good (plotted in blue) is £50,000 and λ for money (plotted in red) is £200,000.

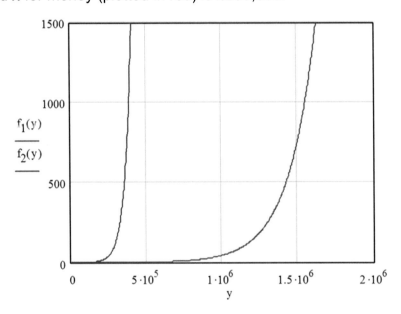

Figure 25. Effect of a Specific Value of λ for a Good

The growth of marginal utility of income arising from consumption of a speculative good for which, $|\lambda_x| < |\lambda|$ increases more steeply as income rises and occurs at a lower income.

The gain in marginal utility of income arising from the emergence of a speculative market for good **x** is then given by:

$$\Delta_{\text{s-c}}\frac{dU_x}{dy} = a_x e^{y/|\lambda_x|} + \frac{a_x y e^{y/|\lambda_x|}}{\lambda_x} - a_x e^{-y/\lambda} + \frac{a_x y e^{-y/\lambda}}{\lambda} \qquad \text{Equation 41}$$

when λ_x has a different absolute value from λ.

This analysis leads to the concepts of a pure good and of quasi-money. Pure goods must always be traded in competitive markets (including monopoly and oligopoly markets) and their marginal utility of income is always given by:

$$\frac{dU_x}{dy} = a_x e^{-y/\lambda} - \frac{a_x y e^{-y/\lambda}}{\lambda}$$

Goods traded in speculative markets have the same element of value as pure goods but also an additional element of value due to their function as a form of quasi-money. Readers may recall our initial discussion of the emergence of speculative markets identified a 'confusion' of these functions as the prime reason for the emergence of such markets. The principle of diminishing marginal utility of income therefore applies to all pure goods and a form of the 'law of diminishing marginal utility' (or something rather like it) is therefore largely rehabilitated. The graph below illustrates this point.

Note that the marginal utility of a competitive or pure good itself (the first derivative of utility or value with respect to demand) is always zero in this system, a consequence of the premise that the level of consumption of a good in the short term is constant for a given state of preference and independent of the prevailing level of price.

The increased value of a speculative good is due to its additional function as a form of quasi-money. This additional value is given by equations 35, 37 or 39. Because they are so different it may not be obvious that equation 40 is in fact a special case of equation 41 in which $\lambda_x = \lambda$. This follows from the definition of the hyperbolic functions sinh and cosh.

The limited domain of demand in a speculative market however places a severe constraint on the extent to which the rich can benefit from this increase in utility. This constraint due to the domain of demand in a speculative market effectively limits the maximum level of income that will increase utility for a consumer even in a speculative market. The maximum level of 'useful' income (for the purposes of 'consumption' of good x) can be calculated as follows. Due to the great complexity of this analysis (the result given below is that achieved after simplification) and to reduce the high probability of error, use has been made of the mathematics package MATHCAD 2001 Professional to obtain the expressions given below. In a speculative market

$$P_x = C_D \cdot \ln(D_x) + \frac{V_x}{D_x}$$

$$U_x = C_D \cdot D_x \cdot \ln(D_x) + V_x$$

Substituting the estimated maximum value of D_x

$$U_x = 2 \cdot V_x \cdot \ln\left(2\frac{V_x}{C_D}\right) + V_x$$

Expanding V_x

$$U_x = 2 \cdot \left(\alpha_x \cdot y_x \cdot \exp\left(\frac{y_x}{\lambda_x}\right) + \beta_x \right) \cdot \ln\left[2 \cdot \frac{\left[\left(\alpha_x \cdot y_x \cdot \exp\left(\frac{y_x}{\lambda_x}\right) \right) + \beta_x \right]}{C_D} \right] + \alpha_x \cdot y_x \cdot \exp\left(\frac{y_x}{\lambda_x}\right) + \beta_x$$

$$\frac{d}{dy_x} U_x = \left[\begin{array}{l} 2 \cdot \lambda_x \cdot \ln(2) + 2 \cdot \lambda_x \cdot \ln\left[\frac{\left[\left(\alpha_x \cdot y_x \cdot \exp\left(\frac{y_x}{\lambda_x}\right) \right) + \beta_x \right]}{C_D} \right] \ldots \\ + \ldots \\ + 2 \cdot y_x \cdot \ln(2) + 2 \cdot y_x \cdot \ln\left[\frac{\left[\left(\alpha_x \cdot y_x \cdot \exp\left(\frac{y_x}{\lambda_x}\right) \right) + \beta_x \right]}{C_D} \right] + 3 \cdot \lambda_x + 3 \cdot y_x \end{array} \right] \cdot \alpha_x \cdot \frac{\exp\left(\frac{y_x}{\lambda_x}\right)}{\lambda_x}$$

For a maximum:

$$\frac{d}{dy_x} U_x = 0$$

Solving for y_x and simplifying:

$$y_x = \left[\lambda_x \cdot W\left[\frac{-1}{2 \cdot \alpha_x \cdot \lambda_x} \cdot \left(-C_D \cdot \exp\left(\frac{-3}{2}\right) + 2 \cdot \beta_x \right) \right] \right]^{-\lambda_x} \qquad\qquad \text{Equation 42}$$

where y_x is the maximum useful income for the purposes of purchasing good x. Note this does not imply an expenditure of y_x on x in a speculative market since consumption depends on preference and priority as well as income
W represents Lambert's W function
ln is the natural or Napierian logarithm $\log_e$
exp is the exponential function
λ_x is the specific value of λ for good x. Note that λ_x is always negative because λ is negative for a speculative good.

There are thus two roots to this equation for y_x. The first can be evaluated analytically having the absolute value of λ_x. The second cannot be obtained analytically because its derivation includes terms in $y.e^{-y/\lambda}$ that have no analytical solution. It can however be evaluated numerically using Lambert's W function (the inverse of $x.e^x$). The definition of Lambert's W function used in MATHCAD is given later in this essay when we consider the partial equilibrium of two speculative goods.

The latter root is often unreal since any W function whose argument is less than - 0.367879441171443 will produce complex roots while at this value the W function evaluates to −1.   When the argument of the W function is greater than - 0.367879441171443 but none the less negative, the W function evaluates to a value greater (less negative) than −1.

Now $$\beta_x \rightarrow 0$$ for conspicuous consumption

and $$0 < \alpha_x \leq 1$$ for all goods purchased from income

Both of these conditions are those most likely to apply to the very rich when buying most goods. It follows that, subject to these constraints, the absolute real value of the second root of equation 40 is always less than or equal to the absolute value of λ_x, since the argument of the W function is always negative and greater than - 0.367879441171443 for real values of the parameters. Therefore the overall maximum real value of useful income from a utility maximising perspective for good x is λ_x. Note that a similar result and the same conclusion will hold even for any exact value of the upper bound of the domain of demand of a speculative good provided jV_x/C_D is substituted for $2V_x/C_D$ (in two places) as the maximum value of D_x where j is the product of the exact upper bound of the domain of demand of good x and the term $C_D/2V_x$. The value of λ for the currency is therefore the maximum value of useful income for all speculative or conspicuous consumption goods. Since all real markets will include many such goods the maximum level of useful income in all markets must be equal to λ. This result will prove very important when we come to consider welfare economics.

This constraint does not apply to any good that is unique because 'unreal' arbitrary prices are sustainable when there is only one supplier. The presence of only one supplier however fails the test of the greatest happiness principle given in equation 30. Furthermore, this constraint does not apply to purchases such as property when made from capital.

The Stock Market

The stock and commodity markets are generally held to be markets that most closely resemble 'ideal' perfectly competitive markets. This is far from the truth. These markets are speculative markets and therefore subject to virtual monopoly conditions and thus a form of monopoly pricing.

In a sellers or bull market because prices are rising 'investors' have a time preference for shares over money or liquidity. λ is therefore negative. If the speculation continues and does not peter out into a terminating saturation phase the terms under the square root sign of the monopoly price equation will eventually equal zero. This condition can be described as a perfect bull market. Any further speculation can be regarded as resulting in no real prices in the market.

Because the stock market's institutions have powerful mechanisms for tracking and publicising prices, multiple prices can never arise. Under these circumstances, since prices are arbitrary, each bargain will be priced erratically and therefore prices will appear to move violently without any apparent justification.   Buyers once again will no longer be able to make rational 'utility maximising' decisions and the bull market will fail and collapse in chaos.

Prices will begin to falter due mainly to a reduction in the number of bargains struck arising from this confusion, and λ will abruptly revert to its normal positive value restoring competitive conditions.   Since virtual monopoly conditions no longer apply, prices will fall suddenly, initiating a buyers' or bear market.

The stock market is therefore in perpetual oscillation between a perfect bull market and the bottom of a bear market. Note that bull markets are always virtual monopoly markets and bear markets are always competitive markets.

Termination of a bear market is always the result of a change of buyer sentiment or preference.   Unlike bull markets, bear market are never self-terminating.   The turning point at the bottom of a bear market is always a saturating decay function.   In contrast termination of a bull market can be either a saturating exponential growth function following a change of sentiment, or an abrupt discontinuity.   The latter is not due to a change of sentiment but to market failure.

The above analysis is of course greatly simplified, considering only overall modium term patterns in the market as a whole or of an individual stock. In reality the movements of the market are complex, in the common not the mathematical sense.   Some shares are rising while others are falling and breaking news can cause frequent and not always complementary shifts in the value of many shares.   These transient effects, which incidentally follow the same rules, are superimposed on the underlying trends in the market as a whole described above.

What makes this analysis new and interesting is that it provides an explanation of and a possible means of predicting cataclysmic changes in the market as a whole or in the fortunes of particular rapidly rising stocks.

The Fashion Market

Fashion is not just confined to clothing and interior design.   Many manufactured goods contain elements of styling and 'life style' choices include things like how and where we live and where we take our holidays.   There are also fashions in such intrinsically unstylish products as the form of mortgage, pension or insurance we take.   Her Majesty the Queen and His Royal Highness Prince Philip's choice to send their sons to Gordonstoun turned an obscure and un-influential school into a fashionable choice for parents. Fashion is thus all pervading.

In the early 1970s there was a fashion for platform soled boots and shoes. Ever higher and more expensive boots and shoes were worn in ever greater quantities. The most famous examples were worn by Gary Glitter and Slade. Many quite serious ankle injuries resulted. By the late 70s the fashion had gone 'out of style', though not as a result of the injuries, and platform shoes could not be sold for love nor money. The style was revived in the late 1990s.

A fashion will have a take-off phase when the new style is adopted by pioneers, a take up phase when it is widely adopted, and a progressive or discontinuous turning point after which the fashion wanes. This consumer behaviour is very similar to the speculative behaviour previously described and is characteristic of a bull market. Fashion is a speculation in style or more particularly in the utility to be derived from being 'in style'.

Speculative markets have 'consumption' functions displaying exponential growth because λ is negative, resulting in inverse demand curves. As we have seen this promotes virtual monopoly conditions allowing suppliers, even though they are numerous, to charge monopoly prices because buyers must limit their search. In the case of fashions affecting a wide group of consumers and not just the rich, suppliers will choose the lower level of oligopoly price allowing a mass market to be established but still providing elevated profits.

Note that even though oligopoly and not monopoly prices are being charged these are dependent on the monopoly level of price as the general pricing solution shows. Thus when virtual monopoly prices become complex, or unreal, so do oligopoly prices.

The end of the bull market for a fashion can arise either from the speculation petering out due to an exogenous change of taste or from failure of the speculative market as discussed above. If market failure causes the end of the bull market λ will abruptly revert to its normal positive value and the demand function reverts to that of a competitive good.

Conditions then re-emerge, characteristic of a bear market. Suppliers can no longer earn their accustomed elevated profits and stop producing the style. Residual stocks of the discontinued style are then sold off cheaply. This is why the poor tend to lag behind the fashion. If market failure ends the bull market it is this failure rather than any declining popularity of the style that destroys the fashion. Indeed, the popularity of the style will not have waned when the market fails.

Currency Markets

Unlike other speculative markets currency markets are almost invariably bull markets because the dynamics of exchange between any two currencies will generally involve one currency rising in value while the other falls. Any equilibrium in the rate of exchange will be acutely unstable and ephemeral. The value of λ_x for one or other of the currencies will therefore almost invariably be negative. A currency market unlike other speculative markets can therefore achieve a perfect bear market as well as a perfect bull market. In fact, a perfect bear market in one currency, with respect to another, is just a perfect bull market in the other. Currency markets unlike other speculative markets can therefore show market failure, and therefore abrupt discontinuities, at both the top and bottom of their price movement cycles not just at the top.

If we return to equations 10 that defined the relationship between a currency's value and the absolute value of money then:

$$\underline{y}\$ = \frac{\pm y_{ABS}}{\lambda\$}$$

and

$$\underline{y}_£ = \frac{\pm y_{ABS}}{\lambda_£}$$

These equations govern the relationship between the value of a currency and the absolute value of money within each economy and not between them. In currency markets two temporarily sustainable dynamic conditions and one unstable equilibrium are possible between each pair of currencies for example:

$$\text{When } \lambda\$ < 0 \text{ AND } \lambda_£ > 0$$

the **$** rises exponentially against the **£**. This condition will continue until there is a change of sentiment or until market failure occurs.

$$\text{When } \lambda\$ > 0 \text{ AND } \lambda_£ < 0$$

the £ rises exponentially against the $ until a change of sentiment or market failure occurs.

$$\text{When } \lambda\$ > 0 \text{ AND } \lambda_£ > 0$$

the two currencies will be in a very unstable transient equilibrium. Any slight upward movement in either currency could trigger a sustained speculation in the rising currency, unless changing market sentiment damps the speculation. If both $\lambda\$$ and $\lambda_£$ are negative the **$** and the **£** will rise together against any third currency for which λ_x is positive.

Hyper-Inflation

Under inflation the price of most and possibly all goods rise over time. Generally consumers with surplus funds will still prefer to maintain liquidity, even though this may entail some loss of value. This is because the rate of inflation needs to be very high indeed before the loss of value from holding money is greater than the depreciation in the value of second hand goods.

If the rate of inflation rises sufficiently, eventually the resale value of more and more classes of goods will provide a better store of value than money. When this happens the function of money as a store of value will fail, not just with respect to goods that attract speculation for those consumers with surplus funds, but for many goods and all consumers.

The parameter λ will then abruptly switch to a negative value not just with respect to a particular good that is an object of speculation but with respect to all goods, or a wide cross section of goods. This will induce a general time preference among consumers for any good over money. These are the conditions for hyper-inflation.

Speculative markets and therefore virtual monopoly conditions will then arise for all goods. Consumers will be forced to limit their search very drastically, allowing all sellers to charge monopoly prices. The advent of universal monopoly pricing in all markets will further aggravate the inflation. As the price of all goods rises more rapidly, ultimately the function of money as a medium of exchange will fail, and widespread barter 'markets' will develop. Consumers will even buy goods they do not need or want just to use in barter if they can persuade anyone to take their money. The purchase of, or exchange of other goods, for cigarettes as a medium of exchange by non-smokers is one example of this.

Eventually the terms under the square root sign of the monopoly pricing equation for all goods can be considered to become negative. There can then be no real prices in the market for any good. Sellers will be forced to unilaterally set arbitrary prices, resulting in multiple prices for essentially similar goods in all markets for all goods. Consumers will not be able to make rational 'utility maximising' decisions concerning any purchase, even though their search times for all goods have been drastically restricted. The function of money as a measure of relative scarcity, utility and value will then fail universally, and the hyper-inflationary currency will collapse in chaos. The most famous example of this is the German hyper-inflation of the early 1920s.

Giffen Goods

For competitive goods the slope of equations 18 and 19 is negative. For a **Giffen** good the slope of the demand function is positive. The conditions for a **Giffen** good are derived from relations 2 and 3 as follows:

$$q_x(\alpha_x e^{-y/\lambda} - \frac{\alpha_x y e^{-y/\lambda}}{\lambda}) > (\alpha_x y e^{-y/\lambda} + \beta_x)\frac{dq_x}{dy}$$

and

$$q_x(\frac{1}{y} - \frac{1}{\lambda}) > (\frac{\beta_x}{\alpha_x y e^{-y/\lambda}} + 1)\frac{dq_x}{dy}$$

now

$$q_x > 0 \quad \text{since demand is positive}$$

$$y \ll \lambda \quad \text{since income is very low for a } \textbf{Giffen} \text{ good}$$

$$\lambda > 0 \quad \text{as normal}$$

and

$$(\frac{1}{y} - \frac{1}{\lambda}) > 0$$

thus

$$D_x(\frac{1}{y} - \frac{1}{\lambda}) > 0$$

now

$$\frac{dq_x}{dy} < 0 \quad \text{since demand falls as income rises for a } \textbf{Giffen} \text{ good}$$

and $\qquad$ $\left(\dfrac{\beta_x}{\alpha_x y e^{-y/\lambda}} + 1 \right) > 0$ $\qquad$ for any consumption at all

therefore $\qquad$ $\beta_x > -\alpha_x y e^{-y/\lambda}$

Now $\qquad$ $y > 0$ $\quad$ and $\quad$ $e^{-y/\lambda} > 0$ $\quad$ always

$\qquad\qquad$ $\beta_x > 0$ $\quad$ and $\quad$ $\alpha_x < 0$ $\quad$ for a **Giffen** good $\quad$ Conditions 2

Conditions 2 give all the conditions for a **Giffen** good. Note these conditions only apply to individuals whose income is low and to goods of high priority. Where a significant proportion of the population have very low incomes a market demand curve for a **Giffen** good can arise. Such conditions generally only occur however in the markets for staple foods, in poor communities, in times of famine. The appearance of a **Giffen** good demand function is rare, depending on very low levels of income. At higher levels of income, the demand function reverts to that of a competitive good.

The form of the consumption function of an individual for a **Giffen** good is as shown in figure 26 below. Note however that though this function appears linear it in fact follows equation 2 where α_x is negative. Note also that consumption can never be negative, except for vomiting which could be considered as a form of 'un-consumption' consequent upon over satiation of short-term needs.

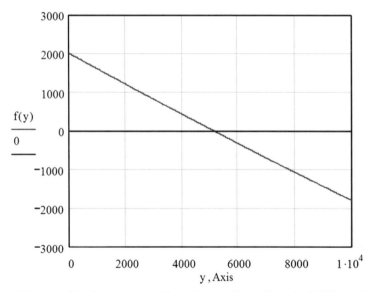

Figure 26 Consumption Function For A Giffen Good

All the conditions established for **Giffen** goods are the opposite of those that apply in speculative markets (except for those conditions that apply in any market). Incomes of participants in the market are low not high, α_x is negative not positive, β_x is positive not negative, λ is positive not negative and the consumption function is a decaying function not a growth function. It seems likely therefore that the demand function of a **Giffen** good is the inverse (in the non-mathematical sense - meaning turned upside down) of that applying in a speculative market. The simplest way of inverting a function (in this sense) is to subtract that function from its maximum value. The effect of 'inverting' the demand function of a speculative market in this way can be seen below. The red line is the speculative demand function and the blue line is the 'inverted' function.

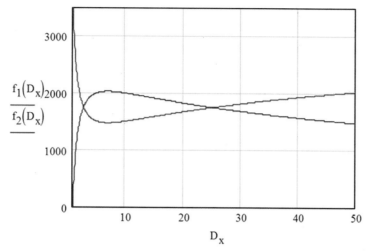

Figure 27. 'Inverting' The Speculative Demand Function

Now the maximum value of the speculative demand function occurs when $D_x = 1$. The term $\log_e D_x$ is then equal to zero and price is equal to V_x (the level of consumption that prevails in a competitive market). If the demand function of a **Giffen** good were the inverse of that of a speculative good then the form of the **Giffen** good demand function would be:

$$P_x = V_x - C_D \log_e D_x - \frac{V_x}{D_x} \qquad \text{Equation 43}$$

This is not so much a mathematical proof or derivation of the form of the **Giffen** good demand function as a suggested model of that function that is consistent with all the conditions given above, a form of economic 'proof' or demonstration. A method of evaluating the parameters of this equation, similar to that used for speculative markets will be considered later. When such parameter values are used, equation 43 should provide a reasonable model of conditions in a **Giffen** good market because the process involves fitting a function of appropriate form to empirical data arising in such a market. Plotting this function produces a graph of the following form:

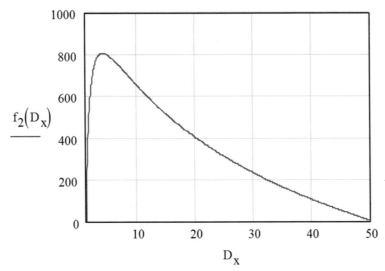

$$f_2(D_x)$$

Figure 28. Demand Function for a Giffen Good

This function certainly behaves in the way we expect the demand function of a **Giffen** good to behave. Inverse demand only occurs at low levels of demand (financed from low levels of income) below the competitive equilibrium level of demand, but at elevated prices above the competitive equilibrium level of price. Inverse demand is ephemeral in a **Giffen** good market. Above the competitive equilibrium level of demand, the demand function reverts to that of a competitive good or at least to a demand function of similar form to that of a competitive good. One aspect of this function however, which is not as is commonly supposed, is that the function is not discontinuous. Though the turning point can be very abrupt under certain circumstance (when direct cost is high relative to the level of consumption) it remains a continuous function.

Just as we can define the domain of demand of a speculative good so we can define its domain for a **Giffen** good. The overall upper bound of demand for a **Giffen** good is the same as that of a speculative good because its demand function 'depends' on that of a speculative good. That is, it is nominally predicted by $2V_x/C_D$. However since the demand function reverts to that of a competitive good (or in our model to a function that behaves similarly to the competitive demand function) above the competitive equilibrium level of demand the upper bound of inverse demand for a **Giffen** good is V_x/C_D. The lower bound is assumed to be 1 because Log_eD_x is undefined when demand is zero.

We can estimate the values of the parameters of the **Giffen** good demand function in a similar way to that used to estimate the parameters of the speculative demand function.

If $\qquad\qquad\qquad\qquad V_G = P_G \cdot D_G$

where $\qquad\qquad\qquad\qquad$ V_G is the **Giffen** good level of consumption
$\qquad\qquad\qquad\qquad\qquad\qquad$ P_G is the price in a **Giffen** good market
and $\qquad\qquad\qquad\qquad\qquad$ D_G is the **Giffen** good level of demand

then
$$\frac{d}{dD_G} V_G = P_G + D_G \frac{d}{dD_G} P_G$$

Let
$$f(D_G) = \frac{d}{dD_G} P_G$$

but
$$V_G = V_C \cdot D_G - C_D \cdot D_G \cdot \ln(D_G) - V_C$$

where
V_C is the competitive level of consumption

therefore
$$\frac{d}{dD_G} V_G = V_C - C_D \cdot \ln(D_G) - C_D$$

Thus, using the MATHCAD solve block:

given
$$P_G + D_G \cdot f(D_G) = V_C - C_D \cdot \ln(D_G) - C_D$$

and
$$P_G = V_C - C_D \cdot \ln(D_G) - \frac{V_C}{D_G}$$

$$\text{Find}(V_C, C_D) \rightarrow \begin{pmatrix} -\dfrac{D_G \cdot P_G - D_G^2 \cdot f(D_G) \cdot \ln(D_G)}{\ln(D_G) - D_G + 1} \\[4mm] -\dfrac{P_G - D_G^2 \cdot f(D_G) + D_G \cdot f(D_G)}{\ln(D_G) - D_G + 1} \end{pmatrix}$$

Therefore
$$V_C = D_G \cdot \frac{\left(-P_G + D_G \cdot f(D_G) \cdot \ln(D_G)\right)}{\left(-D_G + \ln(D_G) + 1\right)}$$

and
$$C_D = \frac{\left(D_G^2 \cdot f(D_G) - P_G - D_G \cdot f(D_G)\right)}{\left(-D_G + \ln(D_G) + 1\right)}$$

If price and demand in a **Giffen** good market are measured on two consecutive days (or weeks or months in a more gradually rising **Giffen** good market):

for
$$P_1 < P_2$$

then

$$P_G = \frac{P_1 + P_2}{2}$$

and for $\quad\quad D_1 < D_2 \quad\quad$ in a **Giffen** good bull market

then

$$D_G = \frac{D_1 + D_2}{2}$$

and

$$f(D_G) = \frac{P_2 - P_1}{D_2 - D_1}$$

Therefore:

$$V_C = \left(\frac{1}{2}\cdot D_1 + \frac{1}{2}\cdot D_2\right)\cdot\frac{\left[\frac{-1}{2}\cdot P_1 - \frac{1}{2}\cdot P_2 + \left(\frac{1}{2}\cdot D_1 + \frac{1}{2}\cdot D_2\right)\cdot\frac{P_2 - P_1}{D_2 - D_1}\cdot\ln\left(\frac{1}{2}\cdot D_1 + \frac{1}{2}\cdot D_2\right)\right]}{\left(\frac{-1}{2}\cdot D_1 - \frac{1}{2}\cdot D_2 + \ln\left(\frac{1}{2}\cdot D_1 + \frac{1}{2}\cdot D_2\right) + 1\right)}$$

and

$$C_D = \frac{\left[\left(\frac{1}{2}\cdot D_1 + \frac{1}{2}\cdot D_2\right)^2\cdot\frac{P_2 - P_1}{D_2 - D_1} - \left(\frac{1}{2}\cdot D_1 + \frac{1}{2}\cdot D_2\right)\cdot\frac{P_2 - P_1}{D_2 - D_1} - \frac{1}{2}\cdot P_1 - \frac{1}{2}\cdot P_2\right]}{\left(\frac{-1}{2}\cdot D_1 - \frac{1}{2}\cdot D_2 + \ln\left(\frac{1}{2}\cdot D_1 + \frac{1}{2}\cdot D_2\right) + 1\right)}$$

In **Giffen** good markets this method of estimation is much less helpful (if we assume the value of C_D is that defined in equation 35) because the reliability of the estimate once again depends on measurements being taken as close as possible to the competitive equilibrium. This occurs in a perfect bull market and for a **Giffen** good that is at the end of the bull market, not at the beginning as was the case for speculative goods. While this may provide a better prediction of the termination of a **Giffen** good market, it may not help prediction throughout the bull market (unless we assume the value of C_D is that which makes equation 35 the closest possible approximation to equation 36). We are therefore likely to have to rely on the prevailing price and volume traded before the **Giffen** good market developed as our basis of estimation, if such records exist. Note that $C_D \approx P_C$ (the competitive equilibrium price) and $V_C = P_C.D_C$ in a competitive market.

Chapter 9. Substitution

This is one of the few uses of comparative static analysis in this essay. It is not that comparative statics is unimportant, it is just that the present essay is concerned with developing fundamental relationships in economics, not with their application.

One process that ordinal utility theory represents very well is the substitution of one good for another. Under ordinal utility theory, changes in preference are exogenous to the model and substitution only occurs as a consequence of a change in the relative price of the two goods. The presumption is that if the price of a good falls consumption of that good is substituted for consumption of others. Substitution between two goods 1 and 2 under ordinal utility theory when the price of good 2 falls by 50% is represented in figure 30. The overall effect can be decomposed in the usual way into an income effect and a substitution effect as shown.

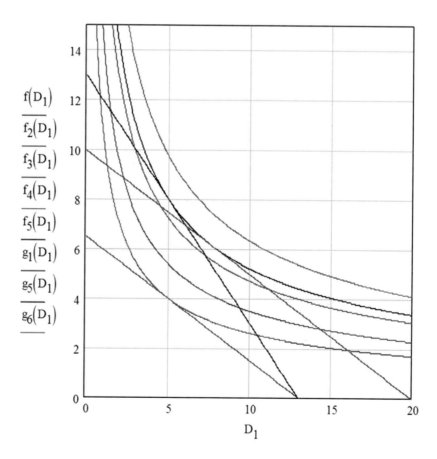

Figure 29. Substitution Due to a Fall in Price Under Ordinal Utility

An increase in income ceteris paribus does not in itself cause substitution of one good for another under ordinal utility theory. The effect of an increase in income is shown in Figure 30. Each of the successive indifference curves from the origin represents expenditure of 1.2 times the preceding curve. These 4 curves are the same curves as similarly coloured curves in figure 29.

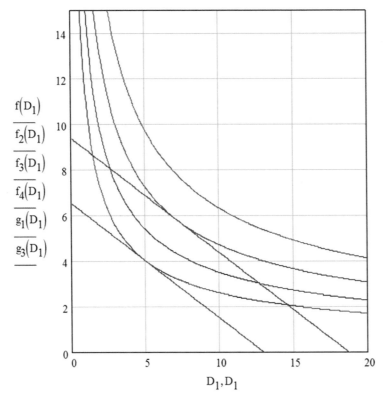

Figure 30. Effect of an Increase In Income Under Ordinal Utility

In both figures 29 and 30 the new budget constraint is the higher budget line.

Under the new theory of value the obverse simplifying assumption applies. The theory generally assumes no substitution takes place between one good and another due to a change in their relative price. At any point in time consumers are assumed to allocate a fixed proportion of their resources to each good in their consumption set. A fall in the price of a good therefore just increases the quantity of that good consumed. This is a simplifying assumption. It is adopted since, except for close substitutes, a fall in the price of one good would in general require substitution for all other goods. This can be simply demonstrated.

If v is a consumer's total consumption of all n goods in his consumption set then:

$$v = \sum_{x=1}^{n} P_x q_x$$

If the price of good r changes then:

$$\frac{dv}{dP_r} = \sum_{x=1}^{r-1}\left(P_x\frac{dq_x}{dP_r} + q_x\frac{dP_x}{dP_r}\right) + q_r + P_r\frac{dq_r}{dP_r} + \sum_{x=r+1}^{n}\left(P_x\frac{dq_x}{dP_r} + q_x\frac{dP_x}{dP_r}\right)$$

Equation 44

Now for a partial equilibrium analysis we may assume:

$$\frac{dv}{dP_r} = 0$$

since an individual's total consumption of all goods will not change merely as a result of a change (particularly a marginal change) in the price of good r.

The quantity of good r consumed is then given by:

$$q_r = -\sum_{x=1}^{r-1} q_x \frac{dP_x}{dP_r} - \sum_{x=r+1}^{n} q_x \frac{dP_x}{dP_r}$$

and the net change in value of good r consumed due to the change in price of good r is given by the terms:

$$P_r \frac{dq_r}{dP_r} = -\sum_{x=1}^{r-1} P_x \frac{dq_x}{dP_r} - \sum_{x=r+1}^{n} P_x \frac{dq_x}{dP_r}$$

which clearly affects all other goods. The effect on the level of consumption of each other good of a marginal change in the price of one good in a partial equilibrium analysis is therefore negligible and can be ignored.

Where goods are close substitutes for one another we simply assume a consumer allocates a fixed budget to a pair of goods instead of to each good individually. This approach is used in equation 45 below and also forms the basis of the definition of virtual homogeneous markets discussed later. This is precisely the same assumption used in ordinal utility theory and incidentally is also a simplifying assumption.

The condition for equilibrium under the two systems can be reduced to identical form. Under the new theory of value if goods A and B are close substitutes then:

$$U_A = P_A q_A$$

Therefore
$$\frac{dU_A}{dq_A} = P_A + q_A \frac{dP_A}{dq_A}$$

Note that this relationship holds irrespective of how P_A is functionally related to q_A save that the function is continuous and differentiable. This applies therefore irrespective of whether a market is competitive and therefore irrespective of the form of the demand function. Thus equation 19 need not apply to this equation.

then
$$\frac{1}{P_A} \frac{dU_A}{dq_A} = 1 + \frac{q_A dP_A}{P_A dq_A}$$

Similarly
$$U_B = P_B q_B$$

$$\frac{dU_B}{dq_B} = P_B + q_B \frac{dP_B}{dq_B}$$

$$\frac{1}{P_B}\frac{dU_B}{dq_B} = 1 + \frac{q_B dP_B}{P_B dq_B}$$

For equilibrium under the new theory of value:

$$\frac{1}{P_A}\frac{dU_A}{dq_A} = \frac{1}{P_B}\frac{dU_B}{dq_B} \qquad \text{Equation 45}$$

therefore

$$\frac{q_A dP_A}{P_A dq_A} = \frac{q_B dP_B}{P_B dq_B}$$

$$\frac{P_B}{P_A} = \frac{q_B dP_B dq_A}{q_A dP_A dq_B} \qquad \text{Equation 46}$$

Now

$$B_{AB} = P_A q_A + P_B q_B \qquad \text{Equation 47}$$

where B_{AB} is the budget constraint for goods A and B

then

$$\frac{dB_{AB}}{dP_A} = q_A + P_A\frac{dq_A}{dP_A} + q_B\frac{dP_B}{dP_A} + P_B\frac{dq_B}{dP_A}$$

now $\dfrac{dB_{AB}}{dP_A} = 0$ since the budget does not change

therefore

$$0 = q_A + P_A\frac{dq_A}{dP_A} + q_B\frac{dP_B}{dP_A} + P_B\frac{dq_B}{dP_A}$$

and

$$P_A\frac{dq_A}{dP_A} = -P_B\frac{dq_B}{dP_A}$$

since if the total value of consumption of goods A and B does not change the increase in expenditure on good A must be equal and opposite to the reduction in expenditure on good B.

thus

$$0 = q_A + q_B\frac{dP_B}{dP_A}$$

$$0 = 1 + \frac{q_B dP_B}{q_A dP_A}$$

therefore

$$\frac{q_B dP_B}{q_A dP_A} = -1$$

Substituting into equation 46 and rearranging gives:

$$\frac{dq_A}{dq_B} = -\frac{P_B}{P_A} \qquad \text{Equation 48}$$

Where $\dfrac{dq_A}{dq_B}$ is the marginal rate of substitution of good A for good B.

This is the equilibrium condition under ordinal utility theory. The fact that this result can be derived from the utility and marginal utility equations 12 and 13 of the new theory of value is a fairly convincing demonstration of its correctness and could be held to be a 'proof of the new theory'. If equations 12 and 13 were incorrect it would not be possible to derive this result.

This analysis applies to all goods, be they competitive, speculative or **Giffen** goods. For competitive goods only, we can deduce a further property of the marginal utility of a good. From the aggregate form of equations 13 and 22:

$$\frac{dU_x}{dD_x} = P_x(1 + \varepsilon_p)$$

Where ε_p is the price elasticity of demand for good x.

but $\varepsilon_p = -1$ from equation 22 and therefore for

competitive goods $\dfrac{dU_x}{dD_x} = 0$

Note this result differs from the result for an individual because individuals are price takers, and therefore the first derivative of price with respect to demand is assumed to be zero since the demand of each individual has a negligible effect on aggregate demand. This does not hold for the market as a whole. It follows that all consumers will be indifferent between the differing quantities of a competitive good **x** that their budget allocation to good x allows them to buy at different prices of **x**. This follows from the fact that their budget allocation to good **x** at any point in time with any given level of income and any given state of preference and priority must be constant and must always be equal to $P_x q_x$.

The new theory of value can also represent substitution of a superior good for an inferior good due to an increase in income, an occurrence only clumsily represented in ordinal utility theory by a change of income inducing an exogenous change of taste. Figures 31, 32 and 33 illustrate the process using the consumption functions of two goods.

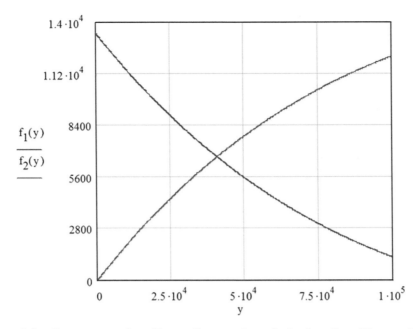

Figure 31. Consumption Functions of an Inferior & a 'Superior' Good

Consumption of the relatively 'superior' good (in red) increases with rising income, while that of the inferior good (in blue) decreases with rising income. If we assume the prices of both goods and the budget remain constant and that $a_2 = - a_1$ throughout, substitution of the superior good for the inferior good as income rises can be represented as shown in figure 32 below.

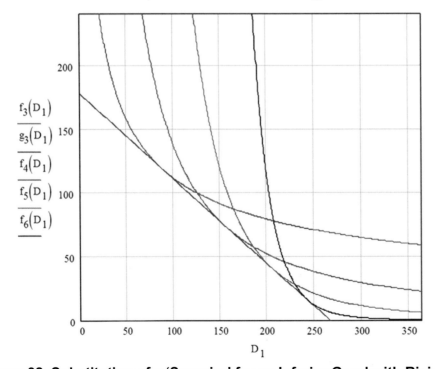

Figure 32. Substitution of a 'Superior' for an Inferior Good with Rising Income

The red indifference curve is plotted for the levels of consumption of each of the two goods that arise at an income of £25,000 and each of the 3 successive curves downwards and to the right of the red curve are plotted for levels of consumption of both goods that arise at incomes of £50,000,

£75,000 and £100,000 respectively, the second and fourth of these coincide with vertical grid lines in figure 32. Note that these conditions only arise when the absolute value (modulus) of preference of the relatively superior and the inferior goods are the same so that total expenditure on both goods remains the same over the region of consumption in which both goods are consumed.

We have been so conditioned by ordinal utility theory that this graph does not seem to represent substitution at all. In fact, that is exactly what is going on. Because the budget and the prices of both goods are constant throughout, the budget line is the same for all 4 equilibrium points. The budget line is in this case a line of constant value or iso-utility curve and locus of all possible equilibrium points under the new theory of value. Therefore despite the fact that each indifference curve is drawn for a different level of income there is no income effect at all in figure 32, just a pure substitution effect with the equilibrium point occurring further downwards and to the right along the budget line at each successive level of income. If the absolute value (modulus) of preference is not the same for the two goods the budget will not be constant and there is an additional income effect, for example the difference between the 4 budget lines in figure 33 below.

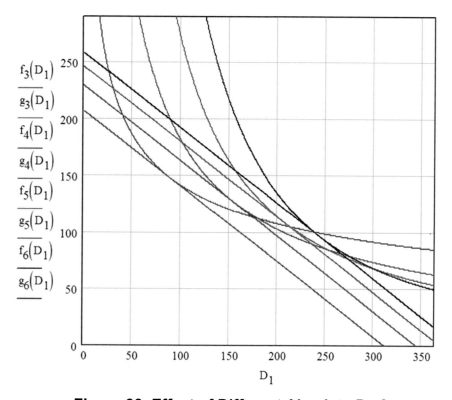

Figure 33. Effect of Different Absolute Preference

Note that these changes in consumption in both examples are endogenous, not exogenous, as is assumed in ordinal utility theory, and do not depend on a change of preference which is also constant throughout in both cases.

Step Function Substitution

While progressive substitution of superior goods for inferior goods as income rises, as described above, is the norm, the new theory of value predicts the existence of step function substitution. Such an effect cannot be represented on an indifference map, though it can be simply represented in the new theory of value using the consumption function. Step function substitution is the 'black hole' of the new theory of value and its existence would be convincing evidence for the correctness of the new theory.

Where goods are mutually exclusive and cannot be consumed simultaneously, substitution of a superior for an inferior good by a particular consumer must take place instantaneously. An example of this is the fashion in hairstyles. In the Edwardian era and the ensuing decade, the fashion was for ladies to wear their hair long and up. In the 1920s a fashion for bobbed hair arose. Since a lady cannot have both long and short hair at the same time, adopting the new fashion required step function substitution. When first introduced bobbed hair was an exclusive fashion. Exclusive high fashion hair stylists are always expensive, so bobbed hair was initially a superior good. Also, at the time, many hairdressers charged by the inch, adding to the initial cost of bobbed hair. Initially, therefore, the high opportunity cost of bobbed hair enabled fashionable ladies to secure an increase in their utility.

Step function substitution can be represented 'schematically' as shown in figure 34. The vertical red and blue lines have been forced apart so that they can be distinguished but they should be interpreted as coincident.

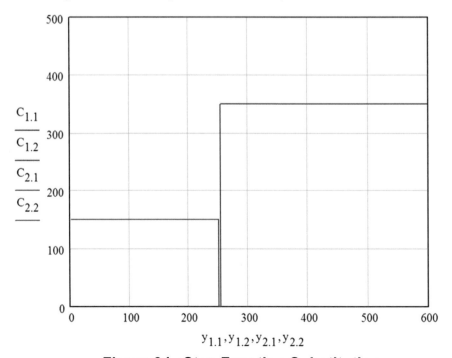

Figure 34. Step Function Substitution

What characterises fashion is rapid uptake of new styles. While an individual's decision to adopt a mutually exclusive new fashion necessarily involves step function substitution even across the market as a whole, the consumption function of a new fashion can be expected to have a steeply rising leading edge approximating a step function. This implies a high positive value of a_2. Mutually exclusive fashions however also imply an equally steeply falling trailing edge of the consumption function of the old style, implying a high negative (more negative) value of a_1. This implies that the overall consumption function of a fashion style can approximate a square wave. This may seem a very strange result but it is perfectly consistent with the model adopted for the new theory of value and, it will be found, with observable reality.

Chapter 10. Partial Equilibrium Between Two Goods in All Markets

All consumers are invariably price takers and obliged to value goods as the market values them. They therefore all subscribe to the same system of values. It follows that, contrary to neoclassical orthodoxy, all consumers must have identical indifference functions. In conventional ordinal utility theory the market demand function for each good is derived by addition of the natural form of each individual's demand function, which in turn is derived from each individual's indifference map. Since all individuals must have identical indifference functions, we should be able to reverse this procedure and derive indifference curves from the market demand function.

As we have derived simple expressions for the demand functions for competitive, speculative and **Giffen** goods, we are therefore in a position to examine the conditions for a partial equilibrium between two goods from all possible combinations of markets. This analysis depends on the form of short-term demand function for competitive goods and is thus prey to any objections concerning the elasticity of that function. It is however a common practice when building very elaborate models, like those developed here, to start from an assumption of the simplest possible building blocks. We have certainly done that. Despite this, some of the mathematics used here is 'new' or more difficult, necessitating the use of the advanced mathematics software MATHCAD 2001 Professional or later versions of MATHCAD.

The most obvious way of deriving equations for indifference curves is to use the result of equation 48 that defines the general condition for the equilibrium point of substitution between two goods. This differential equation can be solved by separation of variables in every case. In some instances, however the resulting algebraic equations cannot be solved analytically and lead to more difficult solutions. Though other methods are available to derive indifference curves in some cases, only this approach based on equation 48 enables indifference curves to be derived in all possible cases.

One other advantage of this method is that when constants of integration are omitted the resulting expression will define a unity or fundamental indifference curve, so called where the curve supports the result $D_2 = 1$ when $D_1 = 1$. Once constants of integration are applied the resulting family of curves therefore provides the basis for defining the set of all possible indifference curves in the region (D_1, $D_2 \geq 1$, D_1, $D_2 \varepsilon R^+$). The lower bound of the domain of D_1 can be assumed to be **1** for all indifference curves because they all contain the term $\log_e D_1$, which is undefined at zero. Since the decision as to which variable to plot on the **x** axis is arbitrary this argument would therefore apply equally to D_2 were this to be the 'independent variable'. The set of all possible combinations of D_1 and D_2 and therefore of all possible indifference curves is thus bounded by the region (D_1, $D_2 \geq 1$, D_1, $D_2 \varepsilon R^+$) or a sub-region thereof.

Whichever method we use, we start this analysis from equation **47** after replacing the suffix A with 1, the suffix B with 2, in preparation for analysis of general equilibrium, and eliminating the suffix AB from the budget B for simplicity.

Hence:

$$B = P_1 D_1 + P_2 D_2 \qquad \text{a form of equation 47}$$

$$\frac{dB}{dD_1} = P_1 + D_1 \frac{dP_1}{dD_1} + D_2 \frac{dP_2}{dD_1} + P_2 \frac{dD_2}{dD_1}$$

but $\quad \dfrac{dB}{dD_1} = 0 \qquad$ since the budget does not change

therefore $\quad P_1 + D_1 \dfrac{dP_1}{dD_1} + D_2 \dfrac{dP_2}{dD_1} + P_2 \dfrac{dD_2}{dD_1} = 0 \qquad$ Equation 49

but $\quad D_1 \dfrac{dP_1}{dD_1} = -D_2 \dfrac{dP_2}{dD_1} \qquad$ Equation 50

since the increase in expenditure on good **1** must be equal and opposite to the reduction of expenditure on good **2** if the budget does not change.

Therefore $\quad \dfrac{dD_2}{dD_1} = -\dfrac{P_1}{P_2} \qquad$ a restatement of $\quad$ Equation 48

Note that this result is entirely general and not dependent on the form of demand functions that applies in any particular case and further, since the equation represents the equilibrium condition, the results obtained, once constants of integration are evaluated, give solutions not just for typical indifference curves but the particular solutions that arise at equilibrium in every case. The only assumptions made, concerning the functions involved are that they are continuous and differentiable. These incidentally are two of the assumptions of ordinal utility theory. We have seen that the three demand functions for competitive, speculative and **Giffen** goods derived earlier are all continuous differentiable functions. The effects of discrete demand functions will be considered later when we review the general equilibrium of real markets.

There are six possible cases: where both goods are competitive **CC**; where one good is competitive and one speculative, a good traded in a speculative market **CS**: where one good is competitive and one a **Giffen** good, a good traded in a **Giffen** good market **CG**; where both goods are speculative **SS**; where one is speculative and one a **Giffen** good **SG** and finally (assuming for the time being that two **Giffen** goods can exist simultaneously) where both goods are **Giffen** goods **GG**. Note that monopoly and oligopoly traded goods depend on the competitive demand functions so they too are represented in this analysis.

The differential equation represented by equation 48 can be solved for any combination of goods as follows:

$$\int P_2\, dD_2 = -\int P_1\, dD_1 \qquad\qquad \text{Equation 51.}$$

This integral equation provides the basis of derivation of all indifference curves. In principle it requires at least one constant of integration unless both sides of the equation were evaluated as definite integrals. We will however generally omit the constant/s of integration initially (except where at least one good is a competitive good, all straight forward cases) to simplify the resulting algebra and correct the omission later.

The CC Case

If both good **1** and good **2** are competitive goods then:

$$P_1 = \frac{V_1}{D_1}$$

and $\qquad\qquad P_2 = \dfrac{V_2}{D_2} \qquad\qquad$ from equation 19

Substituting into equation 51:

$$\int \frac{V_2}{D_2}\, dD_2 = -\int \frac{V_1}{D_1}\, dD_1$$

therefore $\qquad V_2 \cdot \ln(D_2) = -V_1 \cdot \ln(D_1) + k \qquad\qquad$ Equation 52

and solving for **D_2**:

$$D_2 = \exp\left[\frac{\left(-V_1 \cdot \ln(D_1) + k\right)}{V_2}\right] \qquad\qquad \text{Equation 53}$$

This expression is the equation for an indifference curve for two competitive goods and the four steps above constitute the basic method of derivation in all cases. If we omit the constant of integration **k** when plotting this curve, we obtain a graph of the expected strictly convex form.

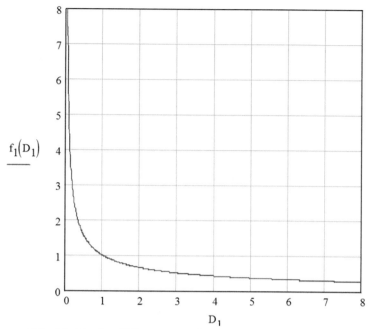

Figure 35. Unity Indifference Curve For Two Competitive Goods

Figure 35 is a particular solution for the **CC** indifference curve because omitting the constant of integration implies the particular value of **k** is zero.

We adopted this approach in this example to demonstrate the fundamental or unity indifference curve described above. Generally, we will be at pains to restrict the calibration of axes of graphs to their correct domain and co-domain or a sub-region thereof. In this case however the graph is plotted with a 'true' origin at (**0, 0**) to illustrate the unity indifference curve. Note that the whole of the indifference curve of figure 35 save the point at (**1, 1**) lies outside the region (**D₁, D₂ ≥ 1, D₁, D₂ ε R⁺**).

If we differentiate equation 53, we obtain the result:

$$\frac{d}{dD_1}D_2 = \frac{-V_1}{D_1 \cdot V_2} \cdot \exp\left[\frac{\left(-V_1 \cdot \ln(D_1) + k\right)}{V_2}\right]$$

which as we might expect simplifies to:

$$\frac{d}{dD_1}D_2 = \frac{-P_1}{P_2}$$

If we solve equation 52 for **k** for any particular pair of equilibrium values of **D₁** and **D₂**, and substitute the resulting value of **k** into equation 53 we obtain the particular solution for that particular equilibrium case.

The methodology used here turns the usual assumptions of mathematics on their head. Generally, analytical solutions are assumed to be exact while numerical solutions are assumed to be approximations. In the present context because the approach is theoretical, the value of parameters may be assumed to be exact. Numerical solutions calculated at extreme or greatest attainable accuracy may also for all practical purposes be assumed to be exact. Very great accuracy is required in evaluating constants of integration to achieve the accuracy and closeness of fit used for the graphs shown below. All calculations are therefore where possible performed to **15** significant figure resolution, the limit of accuracy available from MATHCAD. If you use many fewer significant figures the curves may not quite fit and the error may be apparent even to the naked eye, but even where a lesser accuracy would not suffer in this way, striving for the greatest attainable accuracy provides a means of gauging how good a numerical solution is.

We could evaluate the equilibrium of two goods for the entire market or for an individual. We will choose the latter, the more conventional approach. The levels of demand assumed in our examples will, therefore, all be small, more appropriate to an individual than the market as a whole. Strictly speaking we should also replace V_1 and V_2 with v_1 and v_2 and D_1 and D_2 with q_1 and q_2 in all of the indifference curve equations for an individual to maintain consistency of the symbols used in this paper. For simplicity however, we will retain the symbols used in the 'market' form for all indifference curves in either case. There are two other differences that are explained below.

All the prime examples given below for all six of our original cases will take as parameters:

$$V_1 = 100 \qquad V_2 = 160$$

$$C_1 = 20 \qquad C_2 = 40$$

where C_1 and C_2 are the direct costs of good **1** and good **2** respectively. For the **CC** case equation 52 will be solved and plotted for the equilibrium condition:

$$D_1 = 5 \qquad D_2 = 4$$

We cannot use the same values for D_1 and D_2 in all cases because these values do not necessarily satisfy the conditions for all cases. Generally small integer values have been used in all examples to maintain a relatively uncluttered calibration of both axes in unit quantities so that the equilibrium point can be simply read from the graphs. The system will however work for any values that satisfy the conditions for the case under consideration.

A graph showing equilibrium in the **CC** case with our example data is plotted below:

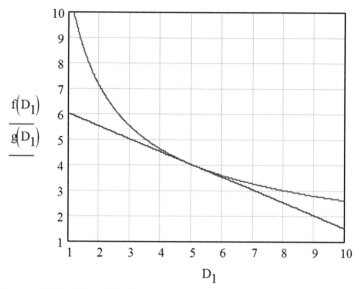

Figure 36. Equilibrium Indifference Curve for the CC Case

The budget function **g(D₁)** shown in blue above is plotted from the equation:

$$D_2 = \frac{B}{P_2} - \frac{P_1}{P_2} \cdot D_1$$

Equation 54

where
$$B = P_1 \cdot D_1 + P_2 \cdot D_2$$

The budget **B** in this case is therefore £260. In all cases **f(D₁)** shown in red represents the indifference curve. The appropriate value of the constant **k** for these conditions in equations 52 and 53 is 382.750889022593.

It may seem strange working this way round from the equilibrium solution to the conditions required to support equilibrium. This approach has been adopted because it is easier to follow the mathematics that way round. This applies particularly in speculative markets where the demand function has no analytical solution for demand in terms of price. Conventional ordinal utility theory assumes the economy works the opposite way-round. That is for a given level of income and a given state of preference the price ratio of the two goods determines the equilibrium level of demand for each good. We will examine this approach to equilibrium when we consider the more difficult **CG**, **SS** and **SG** cases and show that it cannot apply when we investigate general equilibrium.

Because the first derivative of equation 53 naturally simplifies to a slope of – **P₁/P₂** we shall refer to equation 53 and all such equations for indifference curves where only a single constant of integration **k** is used as the natural or market form of the indifference function. The market form of an indifference curve is a kind of average or aggregate indifference curve. It is characteristic of the market form of all indifference functions that at all levels of price for each of the two goods the slope of the indifference curve at equilibrium is always equal to $-P_1/P_2$.

This is generally not the case for an individual because the ratio in which goods are consumed will differ from consumer to consumer, depending on their individual preferences and priorities. Each consumer's equilibrium 'ratio' D_1/D_2 may therefore differ from that of the market as a whole. The natural slope of the fundamental indifference curve at equilibrium for each individual will thus not generally be equal to $-P_1/P_2$. This necessitates the application of two constants of integration k_1 and k_2. k_1 acts to correct the slope by rotating the indifference curve about the individual's equilibrium point to force the slope at that point to equal $-P_1/P_2$ while k_2 acts to correct the position of the curve. The strict form of the CC indifference function for an individual is therefore as given below.

$$D_2 = k_1 \cdot \exp\left[\frac{\left(-V_1 \cdot \ln(D_1)\right)}{V_2}\right] + k_2 \qquad \text{Equation 55}$$

These two constants can be evaluated by equating the first derivative of equation 55 to $-P_1/P_2$ and solving for k_1 and thence solving equation 55 for k_2 for the particular equilibrium values of D_1 and D_2. The appropriate values of the constants k_1 and k_2 for the equilibrium conditions described above are 10.9374541140842 and $3.5527136788005 \times 10^{-15}$ (effectively zero) respectively.

The two sets of equations and their equilibrium points are related to one another as shown in the graph below.

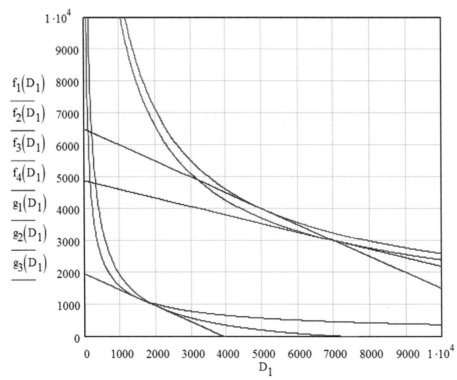

Figure 37. Market & Individual's Equilibrium Points for CC Case

In Figure 37 the two highest indifference curves plotted from equation 53 represent a change in the market equilibrium due to a change of aggregate preference at a constant budget of £260,000. On the highest curve good **1** and good **2** are consumed in the ratio of 5 to 4. On the second curve good **1** and good **2** are consumed in the ratio of 7 to 3. The value of **k** on the highest curve is 2178767.26155795 and on the second curve is 649915.606088335. Note that the slope of the two budget lines at the two market equilibrium points differ because the slope of the budget line simply follows that of the indifference curve. Note also that the two indifference curves do not intersect.

The lowest two curves plotted from equation 55 represent a similar change in the equilibrium for an individual due to the same change of preference with a constant consumer's budget of £78,000 but where the market equilibrium does not change from the position on the highest of the four curves.

The values of k_1 and k_2 on the curve furthest to the left are 200307.09887058 and -779.999999999996 respectively and on the curve second from the left are 115941.331325076 and 0.000000000003689 (effectively zero) respectively. The proportion of total consumption taken by the individual has been greatly exaggerated so that the lowest two curves are not so small that the effect becomes unclear. Otherwise the interaction is entirely representative. The budget line for each of the two equilibrium points for an individual is the same because consumers are invariably price takers and the market equilibrium has not changed. The slope of the budget line is therefore that of the first market equilibrium. Note that two indifference curves for individuals at different levels of preference on a constant budget will necessarily always intersect.

Because the slope of equation 53 at equilibrium can take any negative value we will not complicate the analysis of the 5 other cases by considering both the market and individual equilibrium and simply assume that the slope of the individual consumer's indifference curve at equilibrium is the same as that of the market in our examples. This simplifies the analysis without detracting from the generality of any of the conclusions. We will nevertheless evaluate the two constants required in each case so the result will be typical of any individual's equilibrium.

Just because all consumers have the same indifference functions does not imply they will have identical indifference maps because their incomes, preferences and priorities obviously differ. This affects their allocation of resources to each good and hence, as we shall see, their indifference maps. Now that we have derived the indifference function for an individual in the **CC** case we can demonstrate what we mean by this. The graph below shows a section of the indifference maps of two consumers Mr Brown and Mr Black for two goods. This, like all other graphs in this section, is not a sketch or schematic diagram but an actual graph plotted in MATHCAD from equation 55 using the parametric conditions described below.

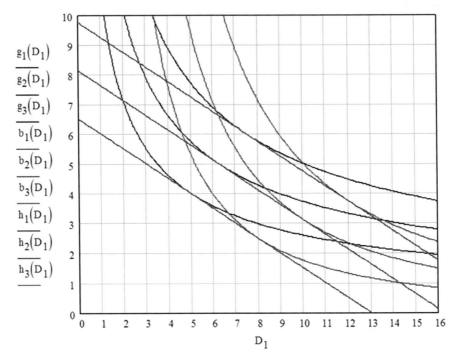

Figure 38. The Indifference Maps of Two Consumers

For simplicity, figure 38 is plotted under the ceteris paribus assumptions that the preferences of both consumers do not change with rising income and that prices are constant throughout.

All six indifference curves shown are plotted using the same indifference function but Mr Brown prefers to consume more of good **1** and less of good **2** (in the ratio of 16 to 5) than Mr Black who prefers the ratio (5 to 4). This affects the parameters V_1, V_2, k_1 and k_2 differently and therefore the two indifference maps are different. Three budget lines are shown in blue. The line nearest to the origin is for a budget of £260 and the two other budget lines represent a 25% and a 50% increase in the budget. The lowest of Mr Black's indifference curves is the same curve as that shown in figure 37. Note that both consumers achieve equilibrium at each level of their budgets but at different levels of demand from one another.

Figure 38 demonstrates that you do not have to assume that each consumer values things differently to generate different indifference maps for each. It also shows that the utility or value each consumer attaches to each good is distinct from his preferences. Utility or value is collectively determined through the market whereas preference is subjective. Exactly the same arguments we have developed here apply to each of the other 5 cases considered below. Indeed, these principles will be even more starkly demonstrated when we consider the **CG**, **SS** and **SG** cases.

Finally to check the accuracy of equilibrium solutions in each case the indifference curve for an individual will also be plotted at the equilibrium point to a resolution of 15 significant figures. In the present case this is achieved by plotting figure 36 over the region D_1 ε (4.99999999999999, 5.00000000000001) and D_2 ε (3.99999999999999, 4.00000000000001). Figure 40 below shows the two graphs of figure 37 plotted to this resolution.

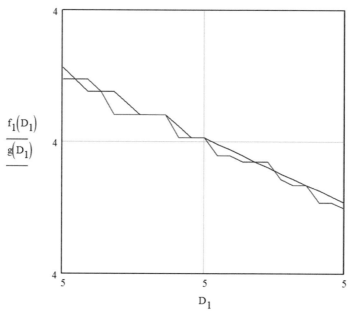

Figure 39. CC Equilibrium Condition at Fine Resolution

At this resolution the two curves, which should effectively be reduced to two (ideally coincident) parallel straight lines, are slightly separated and become very irregular. This is because we are approaching the limit of accuracy of MATHCAD. Note however that the resolution of this graph is 10^{-14}. The process therefore 'finds' the equilibrium condition to a greater accuracy than can ever be required for any real-world case.

The CS Case

If good **2** is a competitive good and good **1** is a speculative good then:

$$P_1 = C_1 \cdot \ln(D_1) + \frac{V_1}{D_1}$$

from equation 35

Substituting into equation 51:

$$\int \frac{V_2}{D_2} \, dD_2 = -\int C_1 \cdot \ln(D_1) + \frac{V_1}{D_1} \, dD_1$$

Equation 56

therefore: $V_2 \cdot \ln(D_2) = -C_1 \cdot D_1 \cdot \ln(D_1) + C_1 \cdot D_1 - V_1 \cdot \ln(D_1) + k$ Equation 57

Solving for **D₂**:

$$D_2 = \exp\left[\frac{\left(-C_1 \cdot D_1 \cdot \ln(D_1) + C_1 \cdot D_1 - V_1 \cdot \ln(D_1) + k\right)}{V_2}\right]$$ Equation 58

Equation 58 gives the market form of the **CS** indifference curve, which differentiates to:

$$\frac{d}{dD_1}D_2 = \frac{-\left(C_1 \cdot \ln(D_1) + \dfrac{V_1}{D_1}\right)}{V_2} \cdot \exp\left[\frac{\left(-C_1 \cdot D_1 \cdot \ln(D_1) + C_1 \cdot D_1 - V_1 \cdot \ln(D_1) + k\right)}{V_2}\right]$$

and simplifies to:

$$\frac{d}{dD_1}D_2 = \frac{-P_1 \cdot D_2}{V_2}$$

and hence:

$$\frac{d}{dD_1}D_2 = \frac{-P_1}{P_2}$$

The form of the **CS** indifference function for an individual is then given by:

$$D_2 = k_1 \cdot \exp\left[\frac{\left(-C_1 \cdot D_1 \cdot \ln(D_1) + C_1 \cdot D_1 - V_1 \cdot \ln(D_1)\right)}{V_2}\right] + k_2$$

     Equation 59

Where good **1** is traded in a speculative bull market the demand for good **1** must exceed the competitive equilibrium level of demand, therefore:

$$D_1 > \frac{V_1}{C_1}$$

Using the parameters given above the equilibrium condition we will investigate in our example complying with this condition is at:

$$D_1 = 6 \qquad\qquad D_2 = 4$$

The value of **B** that satisfies these conditions in equation 54 is £475.011136307367 and the value of **k** that satisfies these conditions in equation 57 and 58 is: 495.994181009355. The appropriate values for **k₁** and **k₂** for these conditions in equation 59 are 22.197143985679 and

5.694011852222417191x10^{-15} (effectively zero) respectively assuming equilibrium for an individual under the same conditions as those of the market as a whole. Figure 40 below shows the equilibrium solution for these conditions.

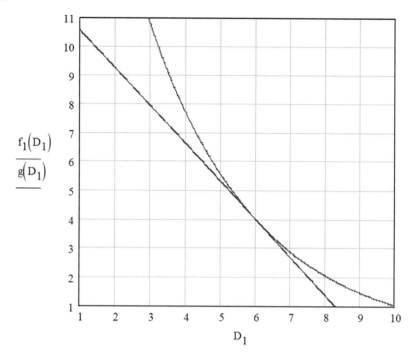

Figure 40. Equilibrium Indifference Curve for the CS Case

Note that the indifference curve for the **CS** case is also convex and that the curve intersects the **x** axis at approximately $2V_1/C_1$, our estimate of the upper bound of the domain of D_1 for a speculative good. As this result is independent of the analysis of domain it appears to confirm that theory. In fact, though this is approximately so in most cases, the upper bound of the domain of D_1 varies with the parameters that apply in each case. This can be demonstrated by a simulation that randomly selects the parameters V_1, V_2, C_1 and C_2 subject to the constraints $V_1 > C_1$ and $V_2 > C_2$ and thence plots the **CS** indifference curve in red and the estimated upper bound of the domain of D_1 as a vertical blue line. In the particular case shown in figure 41 below the equilibrium condition is at D_1 = 6.8513872968071 and D_2 = 6.26655336469412 with V_1, V_2, C_1, C_2 and **k** at £29.9771684137566, £28.603506372381, £6.64997927844524, £4.56447184085846 and 152.302967969176 respectively. Under these conditions, the exact upper bound of the domain of D_1 is 9.81673705232708 and our approximate predictor of this value is 9.01571784168482. A similar simulation and method of calculating the exact upper bound of the domain of D_1 can be developed for each of the other cases with a finite domain of demand. The graph below shows the conditions for this typical example.

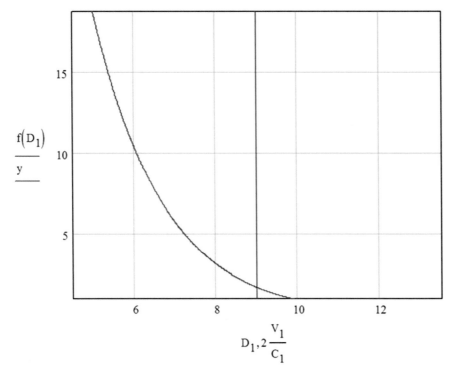

$$D_1, 2\frac{V_1}{C_1}$$

Figure 41. Exact Upper Bound of the Domain of D₁

The 'exact' upper bound of the domain of demand is then given by:

$$D_1 = \frac{2V_1}{C_1}$$

$$D_1 := root\left[exp\left[\frac{\left(-C_1 \cdot D_1 \cdot \ln(D_1) + C_1 \cdot D_1 - V_1 \cdot \ln(D_1) + k\right)}{V_2} \right] - 1, D_1 \right]$$

Equations 60

Note this is a numerical method of solution using the MATHCAD root function starting with an initial first approximation to D₁ of $2V_1/C_1$.

Finally, the accuracy of the equilibrium shown in figure 40 is plotted in figure 42 to 14 decimal place resolution below.

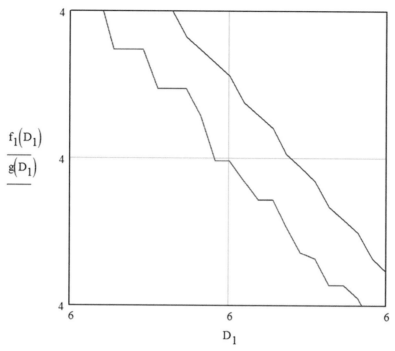

Figure 42. CS Equilibrium Condition to Fine Resolution

The CG Case

If good **1** is a **Giffen** good and good **2** is a competitive good then:

$$P_1 = V_1 - C_1 \cdot \ln(D_1) - \frac{V_1}{D_1} \qquad \text{from equation 43}$$

substituting into equation 51 and integrating gives:

$$\int \frac{V_2}{D_2} dD_2 = -\int \left(V_1 - C_1 \cdot \ln(D_1) - \frac{V_1}{D_1} \right) dD_1 \qquad \text{Equation 61}$$

therefore: $V_2 \cdot \ln(D_2) = -V_1 \cdot D_1 + C_1 \cdot D_1 \cdot \ln(D_1) - C_1 \cdot D_1 + V_1 \cdot \ln(D_1) + k$

Equation 62

and $D_2 = \exp\left[\dfrac{\left(-V_1 \cdot D_1 + C_1 \cdot D_1 \cdot \ln(D_1) - C_1 \cdot D_1 + V_1 \cdot \ln(D_1) + k\right)}{V_2} \right]$ Equation 63

Equation 63 is once again the market form of the **CG** indifference curve.

If we differentiate equation 63, we obtain:

$$\frac{d}{dD_1}D_2 = \frac{\left(-V_1 + C_1 \cdot \ln(D_1) + \dfrac{V_1}{D_1}\right)}{V_2} \cdot \exp\left[\frac{\left(-V_1 \cdot D_1 + C_1 \cdot D_1 \cdot \ln(D_1) - C_1 \cdot D_1 + V_1 \cdot \ln(D_1) + k\right)}{V_2}\right]$$

which once again simplifies to:

$$\frac{d}{dD_1}D_2 = \frac{-P_1}{P_2}$$

The form of the **CG** indifference curve for an individual is then given by:

$$D_2 = k_1 \cdot \exp\left[\frac{\left(-V_1 \cdot D_1 + C_1 \cdot D_1 \cdot \ln(D_1) - C_1 \cdot D_1 + V_1 \cdot \ln(D_1)\right)}{V_2}\right] + k_2$$

Equation 64

Where good **1** is traded in a **Giffen** good bull market, the demand for good **1** must be less than or equal to the competitive equilibrium level of demand.

Therefore:

$$D_1 \leq \frac{V_1}{C_1}$$

Using the parameters given above, the equilibrium condition we will investigate in our example complying with this condition is at:

$$D_1 = 4 \qquad\qquad D_2 = 4$$

The value of **B** that satisfies these conditions in equation 54 is £349.096451110409, the value of **k** that satisfies these conditions in equation 63 is: 452.274112777602 and the values of **k₁** and **k₂** that satisfy these conditions in equation 64 are 16.8898559971546 and $7.105427357601 \times 10^{-15}$ (effectively zero) respectively. Figure 44 below shows the equilibrium solution for these conditions. Note that which variable is plotted on which axis is entirely arbitrary. We have chosen throughout this analysis to assign the simpler demand function (where one of the functions is simpler) to **D₂** and the y-axis for consistency and ease of solution.

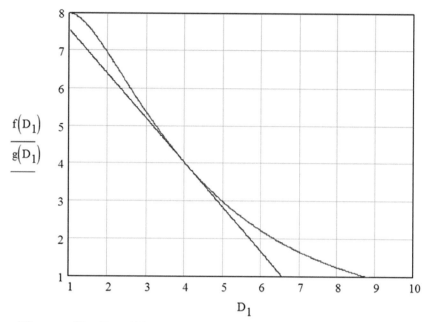

Figure 43. Equilibrium Indifference Curve for the CG Case

The accuracy of this solution for an individual is shown in figure 44 below, a plot of figure 43 to 14 significant figure resolution.

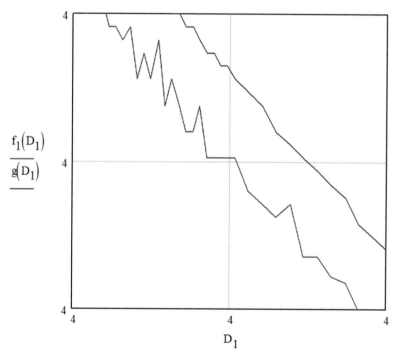

Figure 44. CG Equilibrium Solution at Fine Resolution

Note that the indifference curve for the **CG** case is not strictly convex. We therefore need to consider the conditions for equilibrium rather more precisely.

This curve is of a form that mathematicians describe as di-modal, having a single in-swinging region that is convex to the origin and a single out-swinging region that is 'concave' to the origin.

Orthodox ordinal utility theory assumes that only strictly convex indifference curves can be guaranteed to satisfy the **Karush-Kuhn-Tucker (Kuhn-Tucker**, 1951) optimisation conditions implied by equation 48, and therefore current orthodox theory assumes that a unique equilibrium can only be guaranteed to occur between combinations of goods that give rise to strictly convex indifference curves. Since the only case, that has hitherto been, considered to be guaranteed to produce a strictly convex indifference curve involves substitution between two perfectly competitive goods, current theory has only been able to demonstrate equilibrium in perfectly competitive markets.

We will show that the procedure adopted here 'always' finds the optimum conditions irrespective of whether the indifference curve is strictly convex or not and whether the equilibrium point lies on a convex or concave region of the curve or at a point of inflection between two such regions. Furthermore we will show that the equilibrium solution is necessarily 'always' unique irrespective of whether the goods are traded competitively, speculatively or as **Giffen** goods.

The condition for equilibrium between two goods is generally loosely stated to be defined by equation 48 viz:

$$\frac{d}{dD_1}D_2 = \frac{-P_1}{P_2}$$

In fact, the conditions may be more precisely defined using Boolean logic as:

$$\left(\frac{d}{dD_1}D_2 = \frac{-P_1}{P_2} \text{ AND } \frac{d^2}{dD_1^2}D_2 > 0\right.$$

when the equilibrium point lies on a strictly convex indifference curve or on a strictly convex region of an indifference curve that is not uni-modal,

$$\text{EXCLUSIVE OR} \quad \frac{d}{dD_1}D_2 = \frac{-P_1}{P_2} \text{ AND } \frac{d^2}{dD_1^2}D_2 < 0 \left.\vphantom{\frac{d}{d}}\right)$$

when the equilibrium point lies on a strictly concave indifference curve or on a strictly concave region of an indifference curve that is not uni-modal,

$$\text{EXCLUSIVE OR} \quad \left(\frac{d}{dD_1}D_2 = \frac{-P_1}{P_2} \text{ AND } \frac{d^2}{dD_1^2}D_2 = 0 \right.$$

when the equilibrium point lies at a point of inflection between a convex and concave region of an indifference curve that is not uni-modal, where 'AND' and 'EXCLUSIVE OR' are **Boolean** logic operators.

If we express these relations in terms of optimisation conditions, the first is a utility maximising condition for a given expenditure, the second is a cost minimising condition for a given level of utility, and the last is both a utility maximising and cost minimising condition. An indifference curve that is not uni-modal or that is concave may create a problem for purely numerical methods of solution for which the **Karush-Kuhn-Tucker** conditions were developed, but it does not create a problem for the analytical or 'quasi-analytical' methods used here.

The boundary between the strictly convex region and strictly concave region of the indifference curve in equation 64 can be calculated by taking the second derivative of this function and equating to zero and thence solving for **D₁**.

The second derivative of equation 64 is given by:

$$\frac{d^2}{dD_1^{\,2}}D_2 = \frac{\left(\dfrac{C_1}{D_1} - \dfrac{V_1}{D_1^{\,2}}\right)}{V_2}\cdot\exp\left[\frac{\left(-V_1\cdot D_1 + C_1\cdot D_1\cdot\ln(D_1) - C_1\cdot D_1 + V_1\cdot\ln(D_1) + k\right)}{V_2}\right]\cdots$$

$$+ \frac{\left(-V_1 + C_1\cdot\ln(D_1) + \dfrac{V_1}{D_1}\right)^2}{V_2^{\,2}}\cdot\exp\left[\frac{\left(-V_1\cdot D_1 + C_1\cdot D_1\cdot\ln(D_1) - C_1\cdot D_1 + V_1\cdot\ln(D_1) + k\right)}{V_2}\right]$$

Since $\qquad\dfrac{d^2}{dD_1^{\,2}}D_2 = 0 \qquad$ at the point of inflection

then either, **D₂ = 0**, that cannot be so at any point in the domain of **D₁** or:

$$\frac{\left(\dfrac{C_1}{D_1} - \dfrac{V_1}{D_1^{\,2}}\right)}{V_2} + \frac{\left(-V_1 + C_1\cdot\ln(D_1) + \dfrac{V_1}{D_1}\right)^2}{V_2^{\,2}} = 0$$

This equation has no analytical solution for **D₁** but it can be solved numerically. In our example the value of **D₁** at the point of inflection of the **CG** indifference curve is 2.31455934268683. We can show that MATHCAD has located the point of inflection correctly if we plot the second derivative of equation 64 against **D₁** reproduced below.

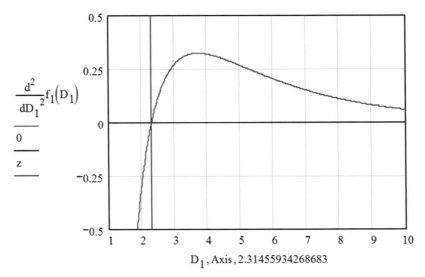

Figure 45. Second Derivative Of CG Indifference Curve

Figure 45 shows the second derivative of the **CG** indifference curve for our example data plotted in red and the calculated point of inflection plotted as a vertical blue line. MATHCAD has clearly identified the point of inflection correctly. We can also illustrate the equilibrium condition at the point of inflection. Figure 46 below shows such an equilibrium for the **CG** indifference curve using our example data.

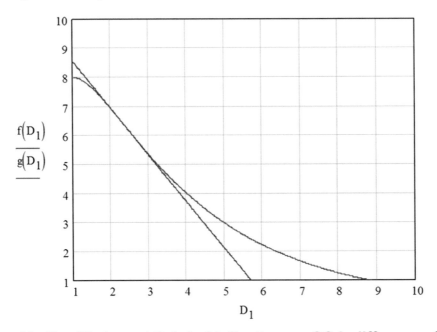

Figure 46. Equilibrium at Point of Inflection on CG Indifference Curve

The equilibrium condition is at **D₁** = 2.3157171558183933276 and **D₂** = 6.4104662144872. The value of **B** to achieve equilibrium is then £369.530339453514. The equilibrium solution still achieves at least 14 decimal place accuracy though reproducing a graph at this resolution is rather messy when the nominal equilibrium point is not at an integer value.

Any equilibrium in which $D_1 <$ 2.31571715558183933276 must therefore lie on the concave region of the curve, for example the point at $D_1 = 1.5$ and $D_2 =$ 7.62287507676139 with a budget **B** of £465.200461055902. A graph at this equilibrium point is shown below.

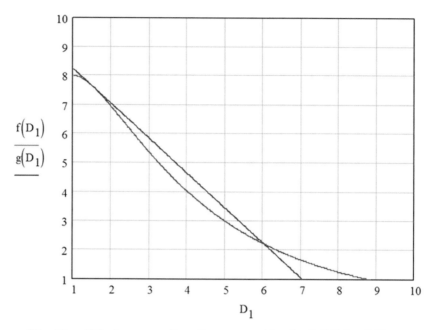

Figure 47. Equilibrium on the Concave Part of CG Indifference Curve

Note that all three equilibrium points and the point of inflection have been plotted for the market equilibrium not those for an individual. The slope of the budget line therefore follows that of the indifference curve, necessitating a different value of the budget in each case. Considering the three graphs figures 43, 46 and 47 we see that in each case the solution is unique. When the equilibrium point is on the convex region of the curve the budget line can only be tangential to the curve at a single point, the equilibrium point. It may intersect the curve at another non-equilibrium point but it can never be tangential at any other point. When the equilibrium point is at the point of inflection of the curve the budget line is both tangential to the curve and intersects the curve at the point of inflection. It cannot then intersect or be tangential to the curve at any other point. When the equilibrium point is on the concave region of the curve the budget line is tangential to the curve at the equilibrium point and cannot be tangential to the curve at any other point. The budget line may intersect the curve at another non-equilibrium point but it can never be tangential to the curve at any other point. Since this set represents all possible cases the equilibrium solution must therefore always be unique on any uni-modal or di-modal indifference curve. We will see later that all but one possible combination of goods with continuous demand functions satisfy these conditions.

Nevertheless the emergence of a di-modal indifference curve appears to create a problem of consumer choice since for almost every point on the concave region of the curve there will be a point on the convex region having the same slope. The issue therefore seems to arise as to which of the two points is the 'preferred' equilibrium point. Note that this issue only arises for indifference curves that are not uni-modal and not for strictly convex or strictly concave indifference curves.

If a consumer had free choice, he would always prefer the point on the convex region of the curve since this would achieve the same level of 'utility' (in the orthodox neo-classical sense) at lower cost. The whole point about the emergence of a **Giffen** good market however is that the consumer does not have free choice but is obliged to increase his level of expenditure to maintain his volume of consumption as best he can. The variable that determines the equilibrium point is the available supply of good **1**. Effective demand is always constrained by the available supply. Since effective demand determines the equilibrium point the solution must always be unique because the indifference curve in the domain of D_1 is one to one.

Strictly speaking this argument only applies, as it stands, to the market solution not to the solution for an individual. However each individual's effective demand for good **1** is constrained by $v_{1G}S_{1G}/V_{1G}$ (where v_{1G} is an individual's consumption of good **1** when it is a **Giffen** good, S_{1G} is the available supply of good **1** when it is a **Giffen** good and V_{1G} is the market consumption of good **1** when it is a **Giffen** good) so the effect is precisely the same for an individual.

In a **Giffen** good market the level of supply is constrained by extreme mischance to be very much lower than the planned level. Furthermore, there is generally little chance in the short term of increasing the supply. The most likely equilibrium condition will therefore tend to be on the concave region of the indifference curve despite consumer preference to the contrary and possibly for quite some time until **Giffen** good conditions begin to ease. Equilibrium will always occur at the lowest point on the curve obtainable within the available supply of good **1**.

A formal analysis of the equilibrium conditions allows a more rigorous treatment of this problem. We can calculate the apparent alternative equilibrium points by solving the equilibrium equation for D_1. From equation 48 the equilibrium equation in the **CG** case is given by:

$$\frac{\left(-V_1 + C_1 \cdot \ln(D_1) + \dfrac{V_1}{D_1}\right)}{V_2} \cdot \exp\left[\frac{\left(-V_1 \cdot D_1 + C_1 \cdot D_1 \cdot \ln(D_1) - C_1 \cdot D_1 + V_1 \cdot \ln(D_1) + k\right)}{V_2}\right] = \frac{-P_1}{P_2}$$

MATHCAD gives two roots (to 14 decimal places) for this equation, using our example data, at 1.48485922714575 and 4. This appears to confirm what economists have long suspected that the equilibrium solution for an indifference curve that is not strictly convex is not unique. We can confirm that MATHCAD has correctly identified both roots by adding P_1/P_2 to both sides of this equation and plotting the resulting expression labelled $d_1(D_1)$ against D_1.

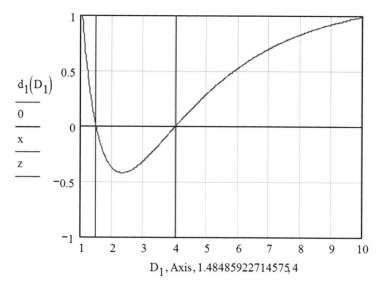

Figure 48. Roots of the CG Equilibrium Equation

Figure 48 clearly shows that MATHCAD has correctly identified both roots of the **CG** equilibrium equation. The apparent dual equilibrium for the **CG** case can be represented as shown below. Figure 49 shows the resulting conditions under the assumptions of ordinal utility theory and figure 50 these conditions under the assumptions of the new theory of value. The two graphs provide a simple but elegant demonstration of the equivalence of the two theories.

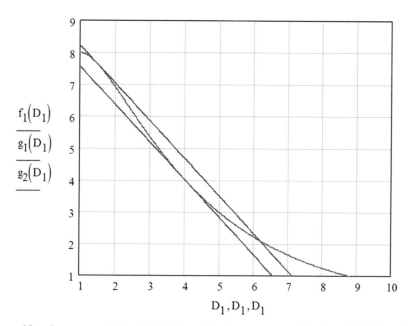

Figure 49. Apparent Dual CG Equilibrium under Ordinal Utility Theory

The second budget line in figure 49 (the higher of the two) implies a budget of £369.530339453514.

Under ordinal utility theory indifference curves are held to represent bundles of goods between which a consumer is indifferent because they are deemed to have equal utility. Indeed in this system indifference curves are more properly described as iso-utility curves. The preference of a consumer for one bundle of goods over another is held to arise because a consumer values the former more dearly than the latter. The line of constant value or utility in this system is the indifference curve. The apparent dual equilibrium is therefore represented by, a single indifference curve and two budget lines.

Clearly if a consumer can obtain two bundles of goods of equal utility but where one is more expensive than the other the cheaper bundle will be preferred. The consumer will therefore always choose the equilibrium solution on the lower budget line unless constrained by the available supply, making the equilibrium solution unique.

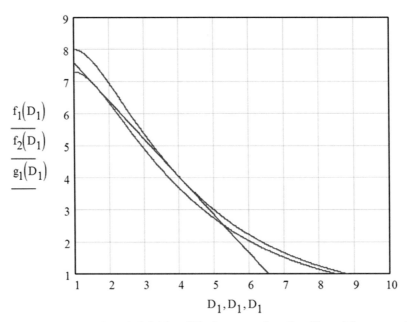

Figure 50. Apparent Dual CG Equilibrium under the New Theory of Value

Under the new theory of value all consumers are obliged to value goods as the market values them because they are price takers. Utility and value are therefore determined collectively through the market while preference is merely subjective. The line of constant value in this system is therefore the budget line. The apparent dual equilibrium is therefore represented by, a single budget line and two indifference curves.

The second indifference curve in figure 50 implies a new value for the constant k of 437.62763601097971321. Note while the two budget lines are parallel in figure 49 because a change of budget affects only the position of the line, the two indifference curves in figure 50 are not 'parallel' because the constant k is part of the argument of an exponential function. Changing the value of k therefore affects both the slope and position of $f_2(D_1)$ not just its position.

Though both bundles of goods at equilibrium have equal value the consumer subjectively prefers the bundle on the higher indifference curve unless constrained by the available supply since for an income less than λ more of any good is always preferred to less. This conjunction of equal value and subjective preference makes the equilibrium solution unique once again. Note that under both systems the equilibrium point lies on the convex region of the equilibrium indifference curve. Contrary to neoclassical orthodoxy both systems therefore produce the same unique solution. It is something of a paradox of partial equilibrium analysis that the uniqueness of the equilibrium solution in both of the two theories depends in some sense on an assumption of the other theory.

We can also demonstrate the uniqueness of the equilibrium solution more rigorously as follows. From equation 48 the **CG** equilibrium equation is given by:

$$\frac{\left(-V_1 + C_1 \cdot \ln(D_1) + \dfrac{V_1}{D_1}\right)}{V_2} \cdot \exp\left[\frac{\left(-V_1 \cdot D_1 + C_1 \cdot D_1 \cdot \ln(D_1)\right) - C_1 \cdot D_1 + V_1 \cdot \ln(D_1) + k}{V_2}\right] = \frac{-P_1}{P_2}$$

but since

$$P_1 = V_1 - C_1 \cdot \ln(D_1) - \frac{V_1}{D_1} \qquad \text{AND} \qquad D_2 = \frac{V_2}{P_2}$$

then − **P₁** appears on both sides of the equation and cancels out. Rearranging and substituting **D₂** for **V₂/P₂** we can simplify the equilibrium equation to:

$$\exp\left[\frac{\left(-V_1 \cdot D_1 + C_1 \cdot D_1 \cdot \ln(D_1)\right) - C_1 \cdot D_1 + V_1 \cdot \ln(D_1) + k}{V_2}\right] = D_2$$

This is exactly the same as equation 63, the expression that defines the market form of the **CG** indifference curve. The reason why the equilibrium solution is unique in the **CG** case is that even though the first derivative of equation 63 is not one to one, the equilibrium equation simplifies to a form identical to the equation to the indifference curve, which is always one to one because the slope of the effective region of all indifference curves in the domain of **D₁** is necessarily always less than or equal to zero. The solution to the **CG** equilibrium equation must therefore 'always' be unique. Though the slope of the **CG** indifference curve may be equal to − **P₁/P₂** at two points on the curve only one of these satisfies the parametric conditions. We will return to discuss this issue more fully when we consider the uniqueness of the more difficult **SS** and **SG** cases.

We can evaluate equilibrium for our example data as follows. With the values of the parameters and the first value of the constant **k** given above, MATHCAD can solve the simplified equilibrium equation for **D₁** and correctly produces a single root of 4 to 14 decimal places. Where an equation has multiple roots MATHCAD will generally find all the roots including unreal roots. Since MATHCAD only returns a single root in this case, we can be fairly certain that, for the parameters given above at least, the solution is unique.

This procedure would however work for any value of **D₂** in its co-domain producing a single real root for **D₁** that will always lie in the domain of **D₁**.

The SS Case

If both goods are speculative goods:

$$P_1 = C_1 \cdot \ln(D_1) + \frac{V_1}{D_1}$$

and

$$P_2 = C_2 \cdot \ln(D_2) + \frac{V_2}{D_2}$$

Substituting into equation 51:

$$\int C_2 \cdot \ln(D_2) + \frac{V_2}{D_2} \, dD_2 = - \int C_1 \cdot \ln(D_1) + \frac{V_1}{D_1} \, dD_1 \qquad \text{Equation 64}$$

Therefore:

$$C_2 \cdot D_2 \cdot \ln(D_2) - C_2 \cdot D_2 + V_2 \cdot \ln(D_2) = -C_1 \cdot D_1 \cdot \ln(D_1) + C_1 \cdot D_1 - V_1 \cdot \ln(D_1) + k$$

Equation 65

This equation can only be solved iteratively for **D₂**. MATHCAD provides a function 'root' that will return an accurate estimate of the dependent variable in such an equation for any value of the independent variable within its domain. Though such iterative solutions will be more or less 'exact' we cannot obtain an exact analytical solution for **D₂**. If however we define an incremental range variable for **D₁** we can define a numeric function for **D₂** using the function root and plot the **SS** indifference curve to a 'similar' accuracy to that achieved using the analytical solutions for earlier cases. The definition is shown below:

$$D_1 = 1, 1.001.. \frac{2V_1}{C_1} \qquad TOL = 10^{-12} \qquad D_2 = \frac{V_2}{C_2}$$

$$f_1(D_1) = \text{root}\left[C_2 \cdot D_2 \cdot \ln(D_2) - C_2 \cdot D_2 + V_2 \cdot \ln(D_2) - \left(-C_1 \cdot D_1 \cdot \ln(D_1) + C_1 \cdot D_1 - V_1 \cdot \ln(D_1) + k \right), D_2 \right]$$

Equations 66

Note the value of **D₂** defined above is only a starting value or initial estimate of **D₂** used in the iterative solution process. We could equally take our equilibrium value of **D₂**. **TOL** is the tolerance function used to specify the accuracy of evaluation. The default setting for **TOL** in MATHCAD is 10^{-15}. At this resolution however MATHCAD is often unable to converge to a solution for the **SS**, **SG** and **GG** indifference curves or their derivatives. Furthermore, graphs of these curves plotted when **TOL** is too low will peter out part way across the plain producing an intermittent ghostly image. To pre-empt this

TOL has been increased to 10^{-12}. Despite this **f₁(D₁)** remains for all practical purposes an exact representation of the **SS** indifference curve and so for simplicity it will be referred to it as the 'true' curve.

For speculative bull markets for both **D₁** and **D₂** demand must exceed the competitive equilibrium level of demand for both goods.

That is: $\qquad$ $D_1 > \dfrac{V_1}{C_1}$ $\qquad$ **AND** $\qquad$ $D_2 > \dfrac{V_2}{C_2}$

We will therefore investigate the equilibrium at:

$$D_1 = 6 \qquad\qquad D_2 = 5$$

complying with these relations.

The value of the constant **k** that satisfies these conditions in equation 65 is 653.584731706448 and the value of **B** that satisfies these conditions in equation 54 is £796.898718794187. Note that this numerical function is effectively the market form of the **SS** indifference curve because it contains only a single constant **k.** The graph below shows the **SS** indifference curve under these conditions. The indifference curve is plotted without a budget line in this case because the curvature is very slight and it is easier to see when the indifference curve is plotted alone.

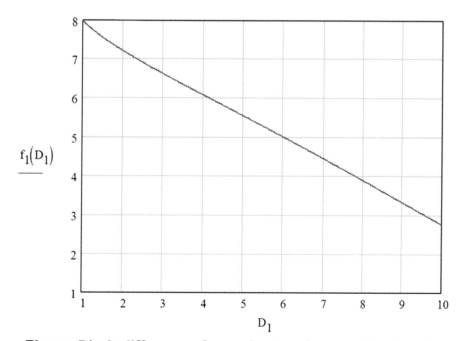

Figure 51. Indifference Curve for two Speculative Goods

This is perhaps the first accurately drawn indifference curve for two speculative goods ever seen. Figure 51 suggests that the **SS** indifference curve might be di-modal. We will investigate this further. Because MATHCAD provides iterative methods of differentiation of numerical functions, we can plot derivatives and calculate approximate points of inflection and roots of the **SS** equilibrium equation even though **f₁(D₁)** is a purely numerical function.

The graph below shows the equilibrium under the conditions given above.

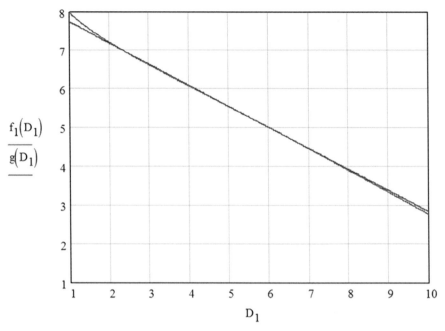

Figure 52. Equilibrium of two Speculative Goods

The accuracy of the solution at equilibrium is shown in figure 53 below a plot of figure 52 to 14 decimal place resolution. Note this also confirms that the intersection of the indifference curve by the budget line is not due to an error of slope on either curve otherwise the two lines would intersect at the equilibrium point even at this resolution.

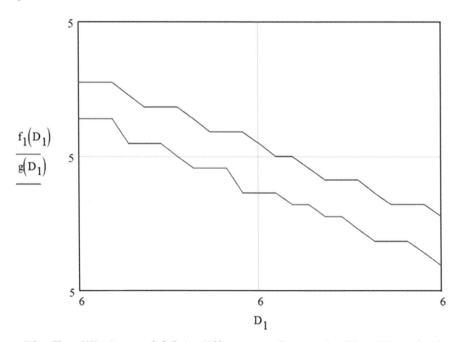

Figure 53. Equilibrium of SS Indifference Curve to Fine Resolution

If we plot the second derivative of $f_1(D_1)$ we obtain the result shown below:

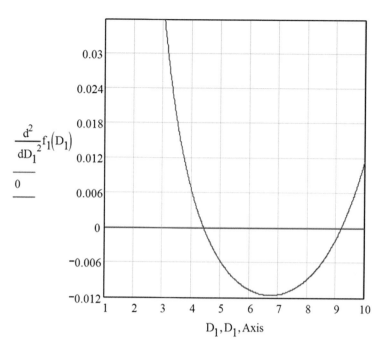

$$\frac{d^2}{dD_1^2}f_1(D_1)$$

$$\frac{0}{\_}$$

Figure 54. Second Derivative of True SS Indifference Curve

Figure 54 shows that the **SS** indifference curve has two points of inflection because the curve intersects the **x** axis twice in the predicted domain of D_1, not just once as we suspected. The **SS** indifference curve is therefore tri-modal. This is not at all obvious from examining figure 51 and makes it feasible that three points of equal slope can occur for certain equilibrium conditions.

Why should the **SS** indifference curve take this form? For any indifference curve for two speculative goods **1** and **2**, at the top left hand end of the curve D_1 will be less than its competitive equilibrium level of demand and therefore good **1** will be in a bear market. D_2 will be greater than its competitive equilibrium level of demand and therefore good **2** will be in a bull market. At the righthand end of the curve the opposite will apply. Because equation 66 was formed by integration of the demand functions of goods **1** and **2** the second derivative of the **SS** indifference curve equation is therefore effectively composed by 'addition' of the slopes of the demand functions of the two goods.

At low levels of demand, near the lower bound of the domain of demand of a speculative good, the slope of its demand functions will be steeply negative. At higher levels of demand, below the competitive equilibrium point, the slope of the demand functions will be slightly negative. Immediately above the competitive equilibrium point the slope of the demand function will be steeply positive. At still higher levels of demand near the upper bound of the domain of demand the slope of the demand function will be slightly positive. Adding the slope of the demand function for good **1** to the slope of a mirror image (left to right about the mid-point of the domain) of good **2** effectively models local fluctuations in slope of the **SS** indifference curve relative to the mean slope of the curve.

The table below shows the additive effect. The first line in row 1 of the table below shows relative changes of slope arising from the demand function for good **1** and the second line in row 1 of the table shows relative changes of slope arising from the demand function for good **2**. In the second row of the table line 3 shows the net additive effect relative to the mean slope of the indifference curve and line 4 shows the consequent local effect on the form of the curve.

1 Steeply Negative 2 Slightly Positive	1 Slightly Negative 2 Steeply Positive	1 Steeply Positive 2 Slightly Negative	1 Slightly Positive 2 Steeply Negative
Negative Dominant Convex	Positive Dominant Concave	Positive Dominant Concave	Negative Dominant Convex

Table 9. Curvature of The SS Indifference Curve

Thus the **SS** indifference curve will take a form which is convex – concave – convex from left to right and thus will have two points of inflection between the three regions.

If we reduce the tolerance to 10^{-11} MATHCAD can evaluate the two points of inflection but only to 5 decimal places as the tables below show.

$$D_1 = 4.401332, 4.4013321..4.401333$$

$$\frac{d^2}{dD_1^2} f_1(D_1) =$$

$7.65497456870627 \cdot 10^{-9}$
$6.38113976884897 \cdot 10^{-9}$
$5.10788232781868 \cdot 10^{-9}$
$3.83396717441475 \cdot 10^{-9}$
0
0
0
0
$-2.53370135440259 \cdot 10^{-9}$
$-3.80688274923335 \cdot 10^{-9}$

The lower point of inflection lies between 4.4013323 and 4.4013328.

$$D_1 \; = \; 9.17659, 9.176591 .. \; 9.1766$$

$$\frac{d^-}{dD_1^2} f_1(D_1) \; =$$

$-4.57936664894415 \cdot 10^{-9}$
$6.37687325165862 \cdot 10^{-9}$
$1.73324125100182 \cdot 10^{-8}$
$2.8288896208353 \cdot 10^{-8}$
$3.9245131514557 \cdot 10^{-8}$
$5.02014817326075 \cdot 10^{-8}$
$6.11574696263497 \cdot 10^{-8}$
$7.21131662648326 \cdot 10^{-8}$
$8.30680401029701 \cdot 10^{-8}$
$9.40258197953735 \cdot 10^{-8}$

The higher point of inflection lies between 9.176590 and 9.176591. MATHCAD cannot converge to a, more accurate, solution in this case because the rate of change of slope is very small along the entire curve making convergence to a solution difficult.

The equilibrium equation for the **SS** indifference curve is given by:

$$\frac{d}{dD_1} f_1(D_1) \; = \; \frac{-P_1}{P_2}$$

If we add **P₁/P₂** to both sides of this equation and plot the resulting expression we obtain the result shown below.

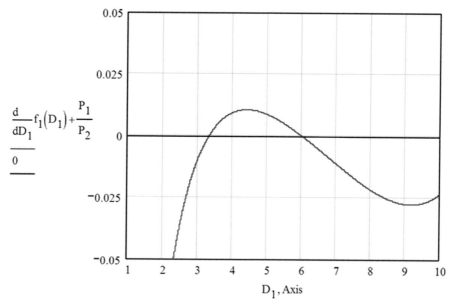

Figure 55. Equilibrium Equation for SS Indifference Curve

This shows that the **SS** equilibrium equation with the parametric conditions given above has two roots in the predicted domain of **D₁** in this particular case. As we might expect MATHCAD confirms a root at 6 to 14 decimal places and can evaluate the second root at 3.31973516841 to 11 decimal places as shown below:

$$D_1 = 3.31973516840, 3.319735168401 .. 3.31973516841$$

$$\frac{d}{dD_1} f_1(D_1) + \frac{P_1}{P_2} =$$

$-1.46438416948058 \cdot 10^{-13}$
$-1.02473585172902 \cdot 10^{-13}$
$-2.76445533131664 \cdot 10^{-14}$
$-3.39728245535298 \cdot 10^{-14}$
$-6.86117829218347 \cdot 10^{-14}$
$-2.74225087082414 \cdot 10^{-14}$
$1.11022302462516 \cdot 10^{-14}$
$5.6843418860808 \cdot 10^{-14}$
$4.47419878923938 \cdot 10^{-14}$
$6.26165785888588 \cdot 10^{-14}$

Equation 66 then gives the associated value of **D₂** to 11 decimal places as 6.44133879858. This result is consistent with the orthodox view that in speculative markets equilibrium solutions need not be unique.

Because we have only a numerical equation defining the **SS** indifference curve we are unable to simplify the **SS** equilibrium equation as we did in the **CG** case. If however we can obtain a close analytical approximation to equation 65 we may be able to demonstrate that a similar simplification applies to the **SS** case resulting in a unique equilibrium solution once again.

The reason why we cannot solve equation 65 analytically is that it contains terms in **D₂**, **Ln(D₂)** and their product. If however we use **Taylor's** approximation to **Ln(D₂)** (about the mid-point of the domain of a speculative good) in the term **C₂.D₂.Ln(D₂)** we should be able to obtain a close approximation to this equation that we may be able to solve using **Lambert's W** function (the inverse of **x.e^x**). The predicted domain of **D₂** is the interval (**1, 2V₂/C₂**). The midpoint of the domain of **D₂** is therefore approximately **V₂/C₂**.

The use of Lambert's W function (Wikipedia, 2019) to solve equations involving exponential functions is not new and has been widely used in the physical and biological sciences particularly by the Nobel Laureates Albert Einstein, Enrico Fermi and Max Planck in nuclear physics but according to Wikipedia at least it has not previously been used in economics or other social sciences.

The first and second order Taylor's approximation to **Ln(D₂)** about **V₂/C₂** in the term **C₂.D₂.Ln(D₂)** are given by:

$$C_2 \cdot D_2 \cdot \ln(D_2) \approx C_2 \cdot D_2 \cdot \left[\ln\left(\frac{V_2}{C_2}\right) + \frac{\left(D_2 - \frac{V_2}{C_2}\right)}{\frac{V_2}{C_2}} \right]$$

and

$$C_2 \cdot D_2 \cdot \ln(D_2) \approx C_2 \cdot D_2 \cdot \left[\ln\left(\frac{V_2}{C_2}\right) + \frac{\left(D_2 - \frac{V_2}{C_2}\right)}{\frac{V_2}{C_2}} - \frac{\left(D_2 - \frac{V_2}{C_2}\right)^2}{2\left(\frac{V_2}{C_2}\right)^2} \right]$$

respectively.

As we might expect the second order (or quadratic) **Taylor's** approximation to **Ln(D₂)** is a better approximation but unfortunately MATHCAD cannot solve equation 66 when either of these two expressions are substituted for the term **C₂.D₂.Ln(D₂)**. If however we substitute **D₂** back for all terms **V₂/C₂** in the denominators of these expressions and substitute the resulting terms for **C₂.D₂.Ln(D₂)** into the left hand side of equation 66 we can form two functions that will be reasonable approximations to the left hand side of equation 66. Note that this procedure, though still quite accurate, involves an approximation to an approximation.

$$t_0(D_2) = C_2 \cdot D_2 \cdot \ln(D_2) - C_2 \cdot D_2 + V_2 \cdot \ln(D_2)$$

$$t_1(D_2) = C_2 \cdot D_2 \cdot \left[\ln\left(\frac{V_2}{C_2}\right) + \frac{\left(D_2 - \frac{V_2}{C_2}\right)}{D_2} \right] - C_2 D_2 + V_2 \cdot \ln(D_2)$$

$$t_2(D_2) = C_2 \cdot D_2 \cdot \left[\ln\left(\frac{V_2}{C_2}\right) + \frac{\left(D_2 - \frac{V_2}{C_2}\right)}{D_2} - \frac{\left(D_2 - \frac{V_2}{C_2}\right)^2}{2 \cdot D_2^2} \right] - C_2 \cdot D_2 + V_2 \cdot \ln(D_2)$$

Where **t₀** is the original form of the left hand side of equation 66 and **t₁** and **t₂** are expressions 'loosely' based on the first and second order **Taylor** approximations to the left hand side of equation 66. If we plot these three functions against **D₂** over the predicted domain of **D₂** for our example data we obtain the result shown below.

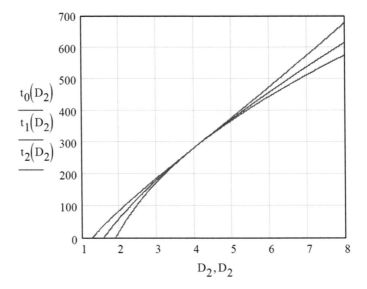

Figure 56. Approximations to the Left-Hand Side of Equation 52

Both **t₁** and **t₂** are good approximations to **t₀** over the domain of **D₂** particularly near the mid-point of the domain, but unusually the expression based on the first order **Taylor's** approximation is better than that based on the second. This is fortunate because MATHCAD cannot solve equation 66 when **t₂** is substituted for **t₀** either. **t₁** uses the second best of the four approximations we have tried for **Ln(D₂)** and is very close to the second order **Taylor's** approximation in a speculative bull market, the region of most interest to us. It also has the advantage of reducing the term **C₂.D₂.Ln(D₂)** to a linear function in **D₂** enabling MATHCAD to solve the resulting equation. If we substitute **t₁** for **t₀** and omit the constant of integration equation 66 becomes:

$$C_2 \cdot D_2 \cdot \left[\ln\left(\frac{V_2}{C_2}\right) + \frac{\left(D_2 - \frac{V_2}{C_2}\right)}{D_2} \right] - C_2 \cdot D_2 + V_2 \cdot \ln(D_2) = -C_1 \cdot D_1 \cdot \ln(D_1) + C_1 \cdot D_1 - V_1 \cdot \ln(D_1)$$

Equation 67

Solving for **D₂** and applying constants of integration:

$$D_2 = \frac{k_1}{\frac{C_2}{V_2} \cdot \ln\left(\frac{V_2}{C_2}\right)} \cdot W\left[\frac{C_2}{V_2} \cdot \ln\left(\frac{V_2}{C_2}\right) \cdot \exp\left[\frac{\left(V_2 - C_1 \cdot D_1 \cdot \ln(D_1) + C_1 \cdot D_1 - V_1 \cdot \ln(D_1)\right)}{V_2} \right] \right] + k_2$$

Equation 68

Where **W** is **Lambert's W** function which we can define in MATHCAD as shown below for all the indifference curves used in this paper involving a **W** function.

$$x = 0, 0.001 .. \frac{2V_1}{C_1} \qquad\qquad y = \frac{V_2}{C_2}$$

$$TOL = 10^{-15} \qquad\qquad W(x) = root\left(y \cdot e^y - x, y\right)$$

$$z = 0, 1 .. ceil\left(\frac{2V_2}{C_2}\right)$$

Where **root** evaluates **Lambert's W** function numerically, **x** is the argument of the **W** function, **TOL** is the tolerance function defining the accuracy of evaluation and **ceil** is the ceiling function that returns the lowest integer that is ≥ **2V2/C2** the estimated upper bound of the domain of demand of a speculative good. Note this definition must be entered as MATHCAD assignments before any **W** functions arise. Because **Lambert's W** function is defined using incremental range variables, it too can be plotted on a graph even though it is a numerical function. Furthermore, it is 'analytically' differentiable. Functions based on the **W** function once defined may therefore be treated in some respects like an analytical function. Procedures using this method may therefore be characterised as 'Quasi-Analytical Approximations'. As far as I am aware this technique is new to mathematics?

To evaluate a particular solution for this approximation to the indifference curve for two speculative goods we need to define two constants even in the market form of the indifference curve, not just one as in earlier cases. There are two reasons for this. Firstly MATHCAD cannot evaluate constants that form part of the argument of a **W** function. Our constants therefore need to be brought outside the **W** function. Secondly, using an approximation to the term **C2.D2.Ln(D2)** introduces a small error in the slope of equation 68 which should be corrected. The constant **k1** acts to correct any error in the slope and **k2** acts to correct the position of equation 68 in the same way as a constant of integration. To calculate **k1** we need to differentiate equation 68. MATHCAD gives the first derivative of equation 68 with respect to **D1** as:

$$\frac{d}{dD_1} D_2 = \frac{-k_1 \cdot \left(C_1 \cdot \ln(D_1) + \dfrac{V_1}{D_1}\right) \cdot W\left[\dfrac{C_2}{V_2} \cdot \ln\left(\dfrac{V_2}{C_2}\right) \cdot \exp\left[\dfrac{\left(V_2 - C_1 \cdot D_1 \cdot \ln(D_1) + C_1 \cdot D_1 - V_1 \cdot \ln(D_1)\right)}{V_2}\right]\right]}{C_2 \cdot \ln\left(\dfrac{V_2}{C_2}\right) \cdot \left[1 + W\left[\dfrac{C_2}{V_2} \cdot \ln\left(\dfrac{V_2}{C_2}\right) \cdot \exp\left[\dfrac{\left(V_2 - C_1 \cdot D_1 \cdot \ln(D_1) + C_1 \cdot D_1 - V_1 \cdot \ln(D_1)\right)}{V_2}\right]\right]\right]}$$

This expression must be equated to − **P1/P2** and solved for **k1** and thence **k2** evaluated from equation 68 for the equilibrium values of **D1** and **D2** to obtain the desired particular solution. Under the conditions specified above the appropriate values for **k1** and **k2** are: 4.4996492911506526691, 3.0964604285278015682 respectively. The resulting approximation to the equilibrium solution labelled **f2(D1)** plotted in brown is shown below plotted together with the true indifference curve **f1(D1)** plotted in red.

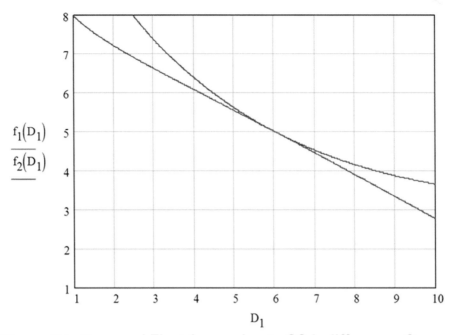

Figure 57. True and First Approximate SS Indifference Curves

Though Equation 68 is accurate to 14 decimal places at the equilibrium point $f_2(D_1)$ is a very poor approximation to the true **SS** indifference curve at all points remote from equilibrium. Furthermore $f_2(D_1)$ is uni-modal and therefore not even of a similar form to the true curve. All in all the approximation is disappointingly inaccurate. However somewhat surprisingly centring our approximation at the upper bound of the domain of demand of D_2 improves the accuracy of approximation as the graph below shows. Figure 58 shows the true indifference curve in red, our approximation centred at V_2/C_2 in brown and an approximation centred at $2V_2/C_2$ labelled $f_3(D_1)$ in black. The improved approximation is calculated from the equation given below.

$$f_3(D_1) := \frac{k_3}{\frac{V_2}{C_2} \cdot \ln\left(\frac{E \cdot V_2}{C_2}\right)} \cdot W\left[\frac{C_2}{V_2} \cdot \ln\left(\frac{E \cdot V_2}{C_2}\right) \cdot \exp\left[\frac{\left(E \cdot V_2 - C_1 \cdot D_1 \cdot \ln(D_1) + C_1 \cdot D_1 - V_1 \cdot \ln(D_1)\right)}{V_2}\right]\right] + k_4$$

Equation 69

Where **E** is an error correcting factor the number of times the centring point of approximation of equation 69 is greater than that of equation 68. **E** also represents a multiple of the midpoint of the domain of D_1. The value of **E** used to plot figure 58 is 2 and the appropriate values of the constants k_3 and k_4 are 2.80673466462011 and 2.60272646845715 respectively.

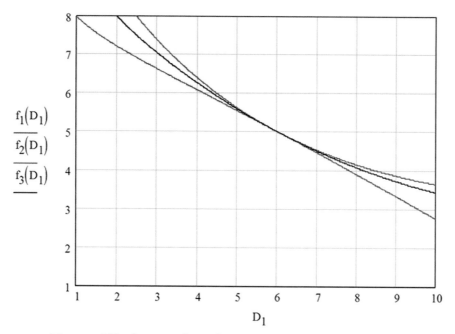

Figure 58. Improving the Accuracy of Approximation

Even more surprisingly when we centre our approximation outside the domain of demand of D_2 (above the upper bound of the domain of demand) we improve the accuracy of approximation still further as figure 59 shows. As the value of **E** is increased $f_3(D_1)$ becomes a closer and closer approximation to the true **SS** indifference curve $f_1(D_1)$. **E,** may indeed have an optimum value. There is however no straightforward way to determine what its optimum value might be because there is a practical limit to this process. When **E** is set to 10.5 the maximum error in $f_3(D_1)$ relative to $f_1(D_1)$ over the domain of D_1 is **1.7%** to 1 decimal place. When **E** is increased above 10.5 MATHCAD crashes.

Increasing the error correction factor has not only improved the accuracy of approximation but it has 'pulled' $f_3(D_1)$ into a tri-modal form so that the analytical approximation now behaves in a way more like the 'true' indifference curve. While setting **E** to 10.5 reduces the error in $f_3(D_1)$ to its practical 'minimum' this may not however be the 'best' setting to adopt because though $f_3(D_1)$ is then tri-modal it is not tri-modal in the predicted domain of D_1. If we limit the maximum setting of **E** to 7.5, we can achieve a significant improvement in the accuracy of $f_3(D_1)$ over that of $f_2(D_1)$ while retaining a quasi-analytical approximation to the true indifference curve that is tri-modal in the domain of D_1. The two graphs below show the 'true' indifference curve and our improved approximation to it with **E** set to 7.5 and the equilibrium of the two curves plotted together. The appropriate values of the constants k_3 and k_4 under these conditions are 1.74842032196841 and − 2 3.61807852975908 respectively. The maximum error in $f_3(D_1)$ relative to $f_1(D_1)$ is then 2.6% to 1 decimal place.

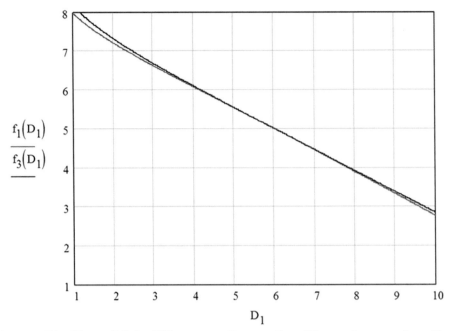

Figure 59. True SS Indifference Curve & a Close Approximation

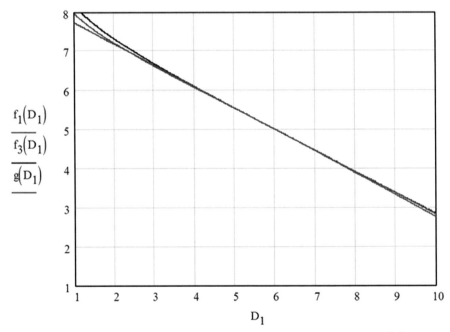

**Figure 60. Equilibrium of True & Approximate SS
Indifference Curves**

Why this error correcting process works is not at all clear but that it does work is certain. We can illustrate the similarity of behaviour of $f_3(D_1)$ to $f_1(D_1)$ in the graphs shown in figures 61 and 62 below.

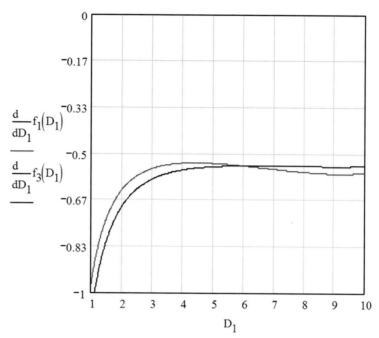

Figure 61. First Derivative of True & Approximate SS Indifference Curve

Figure 61 shows the first derivatives of **f₁(D₁)** and **f₃(D₁)** and figure 62 shows their second derivatives. The slope of **f₃(D₁)** is always very close to that of **f₁(D₁)**.

The second derivative of **f₃(D₁)** is given by MATHCAD as:

$$\frac{d^2}{dD_1^2}f_3(D_1) =$$

$$\frac{k_3 \cdot C_2 \cdot \left(-C_1 \cdot \ln(D_1) - \frac{V_1}{D_1}\right)^2}{V_2^3 \cdot \ln\left(\frac{E \cdot V_2}{C_2}\right)} \cdot \frac{W\left[\frac{C_2}{V_2} \cdot \ln\left(\frac{E \cdot V_2}{C_2}\right) \cdot \exp\left[\frac{(E \cdot V_2 - C_1 \cdot D_1 \cdot \ln(D_1) + C_1 \cdot D_1 - V_1 \cdot \ln(D_1))}{V_2}\right]\right]}{\left[1 + W\left[\frac{C_2}{V_2} \cdot \ln\left(\frac{E \cdot V_2}{C_2}\right) \cdot \exp\left[\frac{(E \cdot V_2 - C_1 \cdot D_1 \cdot \ln(D_1) + C_1 \cdot D_1 - V_1 \cdot \ln(D_1))}{V_2}\right]\right]\right]^2} \dots$$

$$+\frac{-k_3 \cdot C_2 \cdot \left(-C_1 \cdot \ln(D_1) - \frac{V_1}{D_1}\right)^2}{V_2^3 \cdot \ln\left(\frac{E \cdot V_2}{C_2}\right)} \cdot \frac{W\left[\frac{C_2}{V_2} \cdot \ln\left(\frac{E \cdot V_2}{C_2}\right) \cdot \exp\left[\frac{(E \cdot V_2 - C_1 \cdot D_1 \cdot \ln(D_1) + C_1 \cdot D_1 - V_1 \cdot \ln(D_1))}{V_2}\right]\right]^2}{\left[1 + W\left[\frac{C_2}{V_2} \cdot \ln\left(\frac{E \cdot V_2}{C_2}\right) \cdot \exp\left[\frac{(E \cdot V_2 - C_1 \cdot D_1 \cdot \ln(D_1) + C_1 \cdot D_1 - V_1 \cdot \ln(D_1))}{V_2}\right]\right]\right]^3} \dots$$

$$+\frac{k_3 \cdot C_2 \cdot \left(\frac{-C_1}{D_1} + \frac{V_1}{D_1^2}\right)}{V_2^2 \cdot \ln\left(\frac{E \cdot V_2}{C_2}\right)} \cdot \frac{W\left[\frac{C_2}{V_2} \cdot \ln\left(\frac{E \cdot V_2}{C_2}\right) \cdot \exp\left[\frac{(E \cdot V_2 - C_1 \cdot D_1 \cdot \ln(D_1) + C_1 \cdot D_1 - V_1 \cdot \ln(D_1))}{V_2}\right]\right]}{\left[1 + W\left[\frac{C_2}{V_2} \cdot \ln\left(\frac{E \cdot V_2}{C_2}\right) \cdot \exp\left[\frac{(E \cdot V_2 - C_1 \cdot D_1 \cdot \ln(D_1) + C_1 \cdot D_1 - V_1 \cdot \ln(D_1))}{V_2}\right]\right]\right]}$$

Plotting this expression and the second derivative of the numerical function **f₁(D₁)** results in the graph shown below.

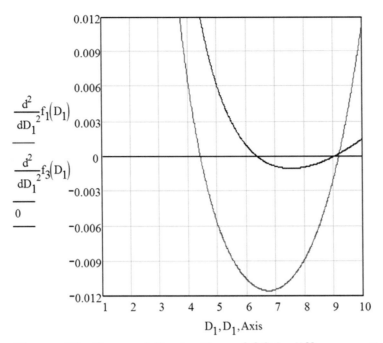

Figure 62. Second Derivative of SS Indifference Curves

While the second derivative is nowhere near as close, $f_3(D_1)$ is clearly tri-modal in the domain of D_1. Note that the greatly expanded scale, on the y-axis of both these and similar graphs shown later, has been adopted to emphasise the differences between the behaviour of the true and approximate indifference curves.

Now that we have a fairly close quasi-analytical approximation to the true **SS** indifference curve that behaves in a similar way as the true curve we are in a position to demonstrate the uniqueness of the all-speculative equilibrium. Assuming our last and most refined approximation to the **SS** indifference curve is a reasonably accurate model of the true curve, then the equilibrium equation for two speculative goods can be represented by:

$$\frac{-k_3 \cdot \left(C_1 \cdot \ln(D_1) + \frac{V_1}{D_1} \right)}{C_2 \cdot \ln\left(\frac{E V_2}{C_2} \right)} \cdot \frac{W \left[\frac{C_2}{V_2} \cdot \ln\left(\frac{E V_2}{C_2} \right) \cdot \exp\left[\frac{\left(E \cdot V_2 - C_1 \cdot D_1 \cdot \ln(D_1) + C_1 \cdot D_1 - V_1 \cdot \ln(D_1) \right)}{V_2} \right] \right]}{\left[1 + W \left[\frac{C_2}{V_2} \cdot \ln\left(\frac{E V_2}{C_2} \right) \cdot \exp\left[\frac{\left(E \cdot V_2 - C_1 \cdot D_1 \cdot \ln(D_1) + C_1 \cdot D_1 - V_1 \cdot \ln(D_1) \right)}{V_2} \right] \right] \right]} = \frac{-P_1}{P_2}$$

This equation has two roots. If we add P_1/P_2 to both sides of this equation and plot the resulting expression we can demonstrate this as the graph below shows.

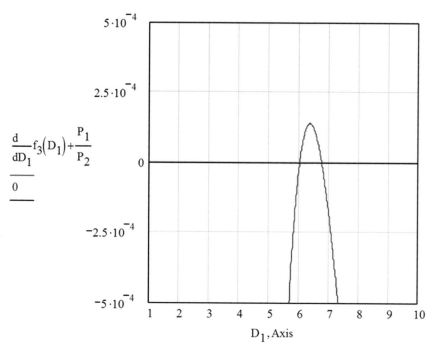

Figure 63. Roots of Un-simplified SS Equilibrium Equation

But

$$P_1 = C_1 \cdot \ln(D_1) + \frac{V_1}{D_1}$$

therefore – **P₁** appears on both sides of the equation and cancels out. The approximate **SS** equilibrium equation therefore simplifies to:

$$\frac{k_3}{C_2 \cdot \ln\left(\frac{E \cdot V_2}{C_2}\right)} \cdot \frac{W\left[\frac{C_2}{V_2} \cdot \ln\left(\frac{E \cdot V_2}{C_2}\right) \cdot \exp\left[\frac{\left(E \cdot V_2 - C_1 \cdot D_1 \cdot \ln(D_1) + C_1 \cdot D_1 - V_1 \cdot \ln(D_1)\right)}{V_2}\right]\right]}{\left[1 + W\left[\frac{C_2}{V_2} \cdot \ln\left(\frac{E \cdot V_2}{C_2}\right) \cdot \exp\left[\frac{\left(E \cdot V_2 - C_1 \cdot D_1 \cdot \ln(D_1) + C_1 \cdot D_1 - V_1 \cdot \ln(D_1)\right)}{V_2}\right]\right]\right]} = \frac{1}{P_2}$$

If we subtract **1/P₂** from both sides of this equation and plot the resulting expression labelled **d₁(D₁)** over the predicted domain of **D₁** in the graph below we can confirm that the simplified approximate **SS** equilibrium equation produces a unique solution once again.

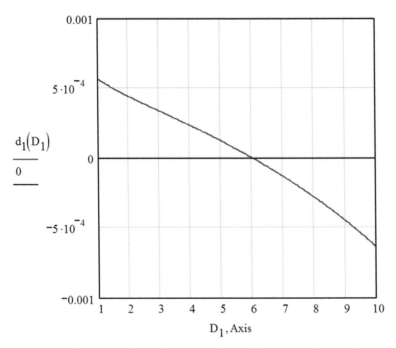

Figure 64. Unique Solution to Simplified SS Equilibrium Equation

MATHCAD can solve the simplified approximate **SS** equilibrium equation producing a single root at 6 to 13 decimal places. Finally, if we plot **f₃(D₁)** together with **f₁(D₁)** and **g(D₁)** at equilibrium to 14 decimal place resolution we obtain the following result:

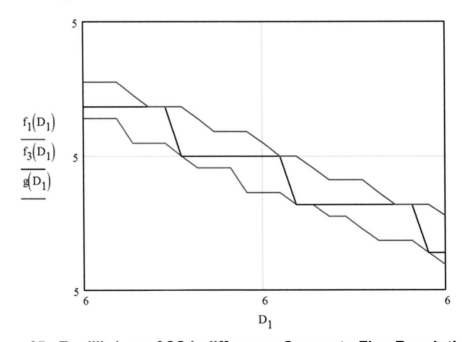

Figure 65. Equilibrium of SS Indifference Curves to Fine Resolution

This shows that using two constants of integration produces a more accurate result and in this case the approximate **SS** indifference curve is more accurate (closer to the nominal equilibrium at the cross hairs) than either the budget line or the true curve. Use of two constants would also improve the accuracy of even the market form of the indifference curves for the cases considered earlier though a second constant is not strictly necessary for those simpler cases.

The SG Case

When good 1 is a **Giffen** good and good 2 a speculative good then:

$$P_1 = V_1 - C_1 \cdot \ln(D_1) - \frac{V_1}{D_1}$$

and

$$P_2 = C_2 \cdot \ln(D_2) + \frac{V_2}{D_2}$$

Substituting into equation 51 gives:

$$\int C_2 \cdot \ln(D_2) + \frac{V_2}{D_2}\, dD_2 = -\int V_1 - C_1 \cdot \ln(D_1) - \frac{V_1}{D_1}\, dD_1 \qquad \text{Equation 70}$$

Therefore:

$$C_2 \cdot D_2 \cdot \ln(D_2) - C_2 \cdot D_2 + V_2 \cdot \ln(D_2) = -V_1 \cdot D_1 + C_1 \cdot D_1 \cdot \ln(D_1) - C_1 \cdot D_1 + V_1 \cdot \ln(D_1) + k$$

$$\text{Equation 71}$$

MATHCAD cannot solve this equation analytically but it can be solved, plotted and differentiated iteratively as a numerical function using the MATHCAD root function, in the same way as we did in deriving the true indifference curve in the **SS** case. The MATHCAD definition of the true numerical form of the market indifference curve is then given by:

$$D_1 = 1, 1.001.. \frac{2V_1}{C_1} \qquad TOL = 10^{-11} \qquad D_2 = \frac{V_2}{C_2}$$

$$f_1(D_1) = \text{root}\left[C_2 \cdot D_2 \cdot \ln(D_2) - C_2 \cdot D_2 + V_2 \cdot \ln(D_2) - \left(-V_1 \cdot D_1 + C_1 \cdot D_1 \cdot \ln(D_1) - C_1 \cdot D_1 + V_1 \cdot \ln(D_1) + k \right), D_2 \right]$$

$$\text{Equations 72}$$

Note the value of D_2 defined above is only a starting value or initial estimate of D_2 used in the iterative solution process and that the tolerance used to specify the resolution of evaluation has been increased to 10^{-11} in this case to 'minimise' convergence problems.

For equilibrium of good 1 in a **Giffen** good bull market and good 2 in a speculative bull market:

$$D_1 \leq \frac{V_1}{C_1} \qquad \text{AND} \qquad D_2 > \frac{V_2}{D_2}$$

If, however we investigate a typical equilibrium complying with these conditions, for example the equilibrium at:

$$D_1 = 4 \qquad\qquad D_2 = 5$$

the resulting graphs of the indifference curves and the budget line are very close together to the left of the equilibrium point, which will make subsequent graphs very difficult to interpret. On this occasion therefore we will investigate the equilibrium of a **Giffen** good at the peak of a perfect bull market and a speculative good in a bull market at:

$$D_1 = 5 \qquad\qquad D_2 = 5$$

Any further increase in demand for the **Giffen** good beyond this point would cause a bear market to develop and the **Giffen** good's demand function to progressively revert to behaviour similar to a competitive good. The process described will however work just as well for any equilibrium conditions within the domain of D_1 and co-domain of D_2

The appropriate value of the constant **k** that satisfies these conditions in equation 71 for the equilibrium values of D_1 and D_2 is 657.510065989456 and the appropriate value of the budget **B** in equation 54 is £720.94379124341. The graph below shows the equilibrium of the true **SG** indifference curve under these conditions:

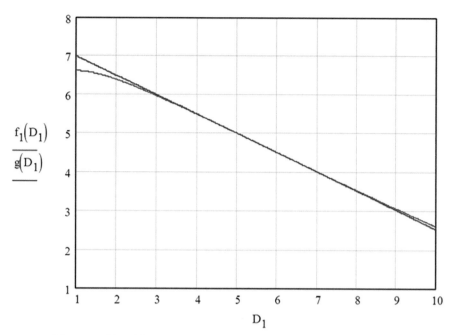

Figure 66. Equilibrium of SG Indifference Curve

Note that the budget line intersects the indifference curve. This suggests that the equilibrium point is at or near a point of inflection indicating that the **SG** indifference curve is not uni-modal. Plotting figure 66 to 14 decimal place resolution confirms that the intersection is not due to an error in slope.

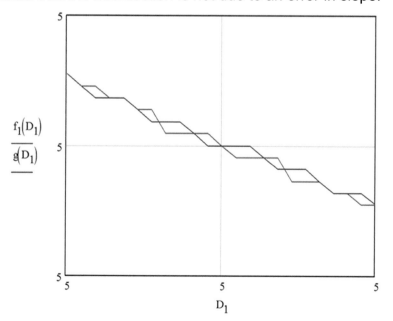

Figure 67. Equilibrium of SG Indifference Curve to Fine Resolution

It may seem strange that an equilibrium should exist between a speculative good and **Giffen** good since speculation and conspicuous consumption are associated with the wealthy and **Giffen** goods with the poor. Perhaps the existence of such an equilibrium explains the poor's love of gambling. Gambling is after all a form of speculation and vice versa.

We can confirm that the true **SG** indifference curve is in fact di-modal by plotting the second derivative of $f_1(D_1)$ numerically as shown below.

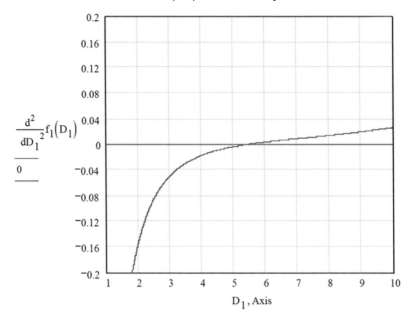

Figure 68. Second Derivative of True SG Indifference Curve

This shows that the true **SG** indifference curve is di-modal because the second derivative of $f_1(D_1)$ intersects the **x** axis once only and it shows that the equilibrium point for our example data is not at the point of inflection. The di-modal form of the **SG** indifference curve arises because a concave region of the curve is induced where a bull market exists for both goods, and a convex region is induced where a bear market exists for both goods.

We can evaluate the point of inflection of the true **SG** indifference curve approximately but only to 5 decimal places as the table below shows

$$D_1 = 5.4977, 5.49771.. 5.4978$$

$$\frac{d^2}{dD_1{}^2}f_1(D_1) =$$

$-1.26575403717199 \cdot 10^{-8}$
$5.92565839135503 \cdot 10^{-8}$
$1.31169855123965 \cdot 10^{-7}$
$2.03084199632508 \cdot 10^{-7}$
$2.74998265456769 \cdot 10^{-7}$
$3.46909964599061 \cdot 10^{-7}$
$4.18822259447985 \cdot 10^{-7}$
$4.90734788697491 \cdot 10^{-7}$
$5.62646073241626 \cdot 10^{-7}$
$6.34557640329912 \cdot 10^{-7}$
$7.06468842025994 \cdot 10^{-7}$

The point of inflection of the true **SG** indifference curve for our example data therefore lies between 5.49770 and 5.49771 and closer to the former because $1.26575403717199 \times 10^{-8} < 5.92565835503 \times 10^{-8}$. Unfortunately, MATHCAD cannot converge to a more accurate solution in this case because the change of slope along much of the true **SG** indifference curve is very slight. Compared with other results in this analysis the point of inflection has been located here very inaccurately. However the resolution is still very much more accurate than might be required for all conceivable real world cases of substitution between a speculative good and a **Giffen** good.

The equilibrium equation for the **SG** indifference curve is given by:

$$\frac{d}{dD_1} f_1(D_1) = \frac{-P_1}{P_2}$$

If we add **P₁/P₂** to both sides of this equation and plot the resulting expression we obtain the result shown below.

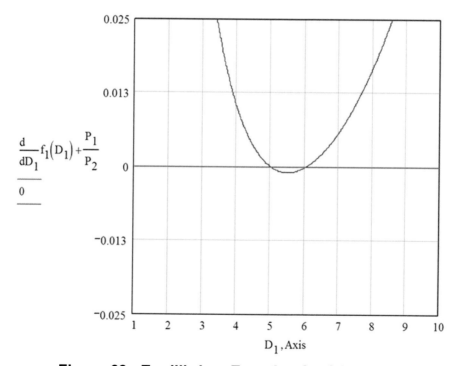

Figure 69. Equilibrium Equation for SG Indifference Curve

This shows that the **SG** equilibrium equation with the parametric conditions given above has two roots in the predicted domain of D_1. As we might expect MATHCAD confirms a root at 5 to 11 decimal places and can evaluate the second root at 6.03779252582 to 11 decimal places as shown below.

$$D_1 = 6.0377925258, \ 6.037792525259 \ .. \ 6.0378925259$$

$$\frac{d}{dD_1} f_1(D_1) + \frac{P_1}{P_2} =$$

-6.79456491070596.10^{-14}
-5.70099523145018.10^{-14}
9.15933995315754.10^{-15}
1.06581410364015.10^{-14}
6.81676937119846.10^{-14}
8.67639293744560.10^{-14}
1.09690034832965.10^{-13}
1.44939615864814.10^{-13}
1.78357328906031.10^{-13}
2.10220729712773.10^{-13}
2.55406806815017.10^{-13}

Equation 72 then gives the associated value of **D$_2$** for the second root to 11 decimal places as 4.48449666624484. This result is once again consistent with the orthodox view that the **SG** equilibrium solution need not be unique.

Because we have only a numerical equation defining the **SG** indifference curve we are unable to simplify this equation as we did in the **CG** case. If, however we can obtain a close analytical approximation to equation 71 we may be able to demonstrate that a similar simplification applies to the **SG** case resulting in a unique solution once again. Since good **2** is a speculative good we can apply the same approximation to **C$_2$.D$_2$.Ln(D$_2$)** as in the **SS** case, then:

$$C_2 \cdot D_2 \cdot \left[\ln\left(\frac{V_2}{C_2}\right) + \frac{\left(D_2 - \frac{V_2}{C_2}\right)}{D_2} \right] - C_2 \cdot D_2 + V_2 \cdot \ln(D_2) = -V_1 \cdot D_1 + C_1 \cdot D_1 \cdot \ln(D_1) - C_1 \cdot D_1 + V_1 \cdot \ln(D_1)$$

Equation 73

Solving for **D$_2$** and applying constants of integration:

$$D_2 = \frac{k_1}{\frac{C_2}{V_2} \cdot \ln\left(\frac{V_2}{C_2}\right)} \cdot W\left[\frac{C_2}{V_2} \cdot \ln\left(\frac{V_2}{C_2}\right) \cdot \exp\left[\frac{\left(V_2 - V_1 \cdot D_1 + C_1 \cdot D_1 \cdot \ln(D_1) - C_1 \cdot D_1 + V_1 \cdot \ln(D_1)\right)}{V_2} \right] \right] + k_2$$

Equation 74

Differentiating with respect to D_1 gives:

$$\frac{d}{dD_1}f_2(D_1) = \frac{-k_1\cdot\left(V_1 - C_1\cdot\ln(D_1) - \frac{V_1}{D_1}\right)}{C_2\cdot\ln\left(\frac{V_2}{C_2}\right)}\cdot\frac{W\left[\frac{C_2}{V_2}\cdot\ln\left(\frac{V_2}{C_2}\right)\cdot\exp\left[\frac{(V_2 - V_1\cdot D_1 + C_1\cdot D_1\cdot\ln(D_1) - C_1\cdot D_1 + V_1\cdot\ln(D_1))}{V_2}\right]\right]}{\left[1 + W\left[\frac{C_2}{V_2}\cdot\ln\left(\frac{V_2}{C_2}\right)\cdot\exp\left[\frac{(V_2 - V_1\cdot D_1 + C_1\cdot D_1\cdot\ln(D_1) - C_1\cdot D_1 + V_1\cdot\ln(D_1))}{V_2}\right]\right]\right]}$$

This expression must be equated to $-$ **P₁/P₂** and solved for **k₁** and thence **k₂** evaluated from equation 74 for the equilibrium values of **D₁** and **D₂** to obtain the particular solution required in the **SG** case.

The values of **k₁** and **k₂** appropriate to the equilibrium conditions given above are: 3.10161994346183 and 4.58463361696320435449 respectively. The graph below shows the true and approximate **SG** indifference curves of equations 72 and 74 plotted together for these conditions.

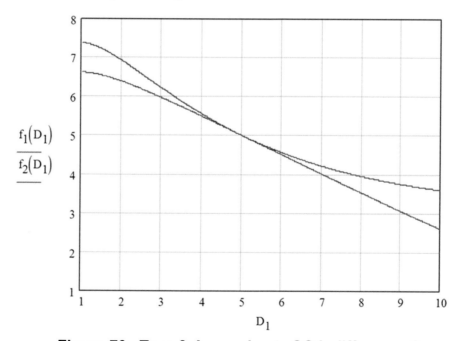

Figure 70. True & Approximate SG Indifference Curves

While **f₂(D₁)** is of the correct di-modal form and is accurate to 14 decimal places at equilibrium it is a poor approximation to **f₁(D₁)** at points remote from equilibrium. Fortunately, all is not lost because we can apply the same error-correcting process used in the **SS** case. Centring our approximation at **E.V₂/C₂** results in the following equation for the **SG** indifference curve:

$$f_3(D_1) = \dfrac{k_3}{\dfrac{C_2}{V_2}\cdot\ln\!\left(E\cdot\dfrac{V_2}{C_2}\right)}\cdot W\!\left[\dfrac{C_2}{V_2}\cdot\ln\!\left(E\cdot\dfrac{V_2}{C_2}\right)\cdot\exp\!\left[\dfrac{\left(E\cdot V_2 - V_1\cdot D_1 + C_1\cdot D_1\cdot\ln(D_1) - C_1\cdot D_1 + V_1\cdot\ln(D_1)\right)}{V_2}\right]\right] + k_4$$

Equation 75

With **E** set to 11.5 the maximum error in **f₃(D₁)** over the predicted domain of **D₁** is reduced to **1%** to 1 decimal place but MATHCAD is unable to converge to a solution for all values of **D₁** below 2 and increasing the tolerance function still further does not prevent this. At higher values of **E**, MATHCAD crashes. When **E** is set to 11.4 the maximum error in **f₃(D₁)** over the domain of **D₁** is less than 1.1% without preventing convergence at any point in the domain of **D₁**.

The appropriate values of **k₃** and **k₄** in this equation for these conditions with the parameters given above are 1.7800621189045 and - 7.91585845540231 respectively. When we plot all three indifference, curve functions together, with the additional function **f₃(D₁)** plotted in black, the result is as shown below. However, because **f₃(D₁)** is such a close approximation to **f₁(D₁)** that the two functions cannot be easily distinguished, a small increment of 0.05 has therefore been subtracted from **f₁(D₁)** so the two functions become distinct.

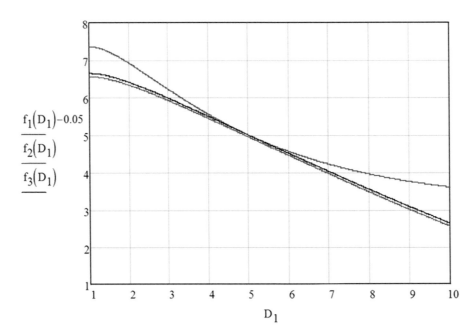

Figure 71. Improving the Accuracy of Approximation of SG Curve

Clearly **f₃(D₁)** is now a very good model of **f₁(D₁)** in the domain of **D₁** for the **SG** case and what is more the approximation behaves in a very similar way to the true indifference curve as the graphs of the first and second derivative of **f₁(D₁)** and **f₃(D₁)** show. Figure 72 shows the first derivatives of **f₁(D₁)** and **f₃(D₁)**.

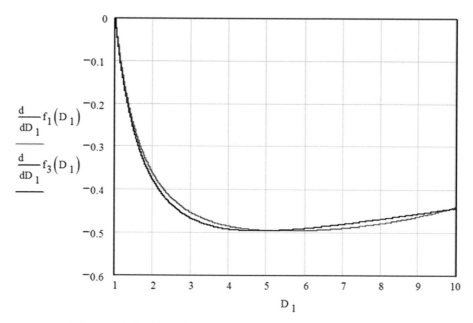

Figure 72. The First Derivative of SG Indifference Curves

MATHCAD gives the second derivative of $f_3(D_1)$ as:

$$\frac{d^2}{dD_1^2}f_3(D_1) = \frac{k_3 \cdot \left(V_1 - C_1 \cdot \ln(D_1) - \frac{V_1}{D_1}\right)^2}{V_2 \cdot C_2 \cdot \ln\left(E \cdot \frac{V_2}{C_2}\right)} \cdot \frac{W\left[\frac{1}{V_2} \cdot C_2 \cdot \ln\left(E \cdot \frac{V_2}{C_2}\right) \cdot \exp\left[\frac{-\left(-E \cdot V_2 + V_1 \cdot D_1 - C_1 \cdot D_1 \cdot \ln(D_1) + C_1 \cdot D_1 - V_1 \cdot \ln(D_1)\right)}{V_2}\right]\right]}{\left[1 + W\left[\frac{1}{V_2} \cdot C_2 \cdot \ln\left(E \cdot \frac{V_2}{C_2}\right) \cdot \exp\left[\frac{-\left(-E \cdot V_2 + V_1 \cdot D_1 - C_1 \cdot D_1 \cdot \ln(D_1) + C_1 \cdot D_1 - V_1 \cdot \ln(D_1)\right)}{V_2}\right]\right]\right]^2} \ldots$$

$$+ \frac{-k_3 \cdot \left(V_1 - C_1 \cdot \ln(D_1) - \frac{V_1}{D_1}\right)^2}{V_2 \cdot C_2 \cdot \ln\left(E \cdot \frac{V_2}{C_2}\right)} \cdot \frac{W\left[\frac{1}{V_2} \cdot C_2 \cdot \ln\left(E \cdot \frac{V_2}{C_2}\right) \cdot \exp\left[\frac{-\left(-E \cdot V_2 + V_1 \cdot D_1 - C_1 \cdot D_1 \cdot \ln(D_1) + C_1 \cdot D_1 - V_1 \cdot \ln(D_1)\right)}{V_2}\right]\right]^2}{\left[1 + W\left[\frac{1}{V_2} \cdot C_2 \cdot \ln\left(E \cdot \frac{V_2}{C_2}\right) \cdot \exp\left[\frac{-\left(-E \cdot V_2 + V_1 \cdot D_1 - C_1 \cdot D_1 \cdot \ln(D_1) + C_1 \cdot D_1 - V_1 \cdot \ln(D_1)\right)}{V_2}\right]\right]\right]^3} \ldots$$

$$+ \frac{-k_3 \cdot \left(\frac{-C_1}{D_1} + \frac{V_1}{D_1^2}\right)}{C_2 \cdot \ln\left(E \cdot \frac{V_2}{C_2}\right)} \cdot \frac{W\left[\frac{1}{V_2} \cdot C_2 \cdot \ln\left(E \cdot \frac{V_2}{C_2}\right) \cdot \exp\left[\frac{-\left(-E \cdot V_2 + V_1 \cdot D_1 - C_1 \cdot D_1 \cdot \ln(D_1) + C_1 \cdot D_1 - V_1 \cdot \ln(D_1)\right)}{V_2}\right]\right]}{\left[1 + W\left[\frac{1}{V_2} \cdot C_2 \cdot \ln\left(E \cdot \frac{V_2}{C_2}\right) \cdot \exp\left[\frac{-\left(-E \cdot V_2 + V_1 \cdot D_1 - C_1 \cdot D_1 \cdot \ln(D_1) + C_1 \cdot D_1 - V_1 \cdot \ln(D_1)\right)}{V_2}\right]\right]\right]}$$

Figure 73 plots the second derivatives of $f_1(D_1)$ and $f_3(D_1)$:

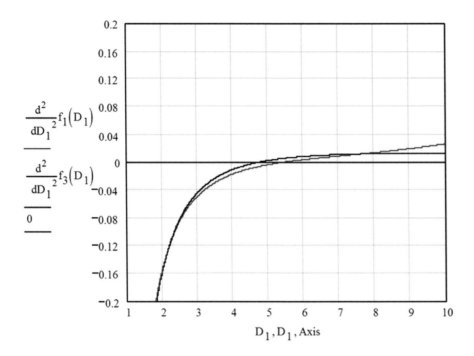

Figure 73. Second Derivative of SG Indifference Curves

Both the first and second derivatives of $f_3(D_1)$ are very close to those of $f_1(D_1)$ throughout the predicted domain of D_1. Note that these graphs are drawn with **E** set to 10.5. MATHCAD crashes when producing these graphs at higher values of **E**. The appropriate values of k_3 and k_4 under these conditions are 1.7800621189045 and - 7.91585845540231 respectively.

Now that we have a very close quasi-analytical approximation to the true **SG** indifference curve that behaves in much the same way as the **true** curve we are in a position to show the uniqueness of the **SG** equilibrium. Assuming our last and most refined approximation to the **SG** indifference curve is an accurate model of the true curve then the equilibrium equation for substitution between a speculative good and a **Giffen** good can be represented by:

$$\frac{k_3 \cdot \left(-V_1 + C_1 \cdot \ln(D_1) + \dfrac{V_1}{D_1}\right)}{C_2 \cdot \ln\left(E \cdot \dfrac{V_2}{C_2}\right)} \cdot \frac{W\left[\dfrac{C_2}{V_2} \cdot \ln\left(E \cdot \dfrac{V_2}{C_2}\right) \cdot \exp\left[\dfrac{\left(E \cdot V_2 - V_1 \cdot D_1 + C_1 \cdot D_1 \cdot \ln(D_1) - C_1 \cdot D_1 + V_1 \cdot \ln(D_1)\right)}{V_2}\right]\right]}{\left[1 + W\left[\dfrac{C_2}{V_2} \cdot \ln\left(E \cdot \dfrac{V_2}{C_2}\right) \cdot \exp\left[\dfrac{\left(E \cdot V_2 - V_1 \cdot D_1 + C_1 \cdot D_1 \cdot \ln(D_1) - C_1 \cdot D_1 + V_1 \cdot \ln(D_1)\right)}{V_2}\right]\right]\right]} = \frac{-P_1}{P_2}$$

This equation has two roots. If we add P_1/P_2 to both sides of this equation and plot the resulting expression we can demonstrate this, as figure 74 below shows. Note that even when **E** is set to 10.5 MATHCAD cannot converge to a solution across the entire displayed region of this curve. Figure 74 is therefore drawn with **E** set to 9.5. The appropriate values of the constants k_3 and k_4 under these conditions are 1.76591117590892 and - 6.44341536552827 respectively.

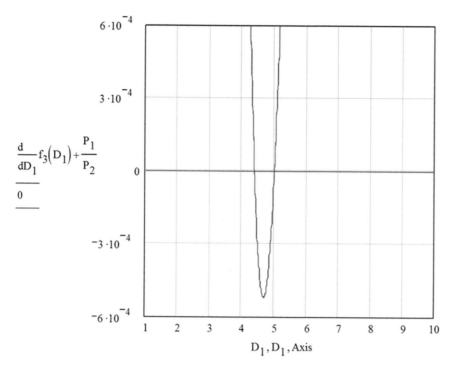

$$\frac{d}{dD_1}f_3(D_1)+\frac{P_1}{P_2}$$

$$\overline{0}$$

Figure 74. Roots of Un-simplified SG Indifference Curve

But
$$P_1 = V_1 - C_1\cdot\ln(D_1) - \frac{V_1}{D_1}$$

therefore – **P₁** appears on both sides of the equilibrium equation and cancels out. The **SG** equilibrium equation then simplifies to:

$$\frac{k_3}{C_2\cdot\ln\left(E\cdot\frac{V_2}{C_2}\right)}\cdot\frac{W\left[\frac{C_2}{V_2}\cdot\ln\left(E\cdot\frac{V_2}{C_2}\right)\cdot\exp\left[\frac{\left(E\cdot V_2 - V_1\cdot D_1 + C_1\cdot D_1\cdot\ln(D_1) - C_1\cdot D_1 + V_1\cdot\ln(D_1)\right)}{V_2}\right]\right]}{\left[1 + W\left[\frac{C_2}{V_2}\cdot\ln\left(E\cdot\frac{V_2}{C_2}\right)\cdot\exp\left[\frac{\left(E\cdot V_2 - V_1\cdot D_1 + C_1\cdot D_1\cdot\ln(D_1) - C_1\cdot D_1 + V_1\cdot\ln(D_1)\right)}{V_2}\right]\right]\right]} = \frac{1}{P_2}$$

If we subtract **1/P₂** from both sides of this equation and plot the resulting expression labelled **d₁(D₁)** over the predicted domain of **D₁** in the graph below, we can confirm that the simplified **SG** equilibrium equation produces a unique solution once again. Although MATHCAD can converge to a solution to **d₁(D₁)** across the predicted domain of **D₁** even when **E** is set to 11.4 this graph has been drawn assuming **E** is set to 9.5 for consistency with figure 74.

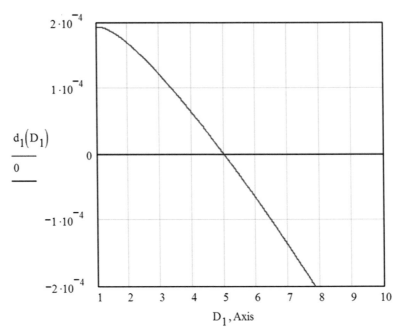

Figure 75. Unique Solution of Simplified SG Equilibrium Equation

MATHCAD can solve the simplified **SG** equilibrium equation producing a single root of 5 to 13 decimal places. Finally, if we plot $f_3(D_1)$ together with $f_1(D_1)$ and $g(D_1)$ at equilibrium to 14 decimal place resolution we obtain the result shown below.

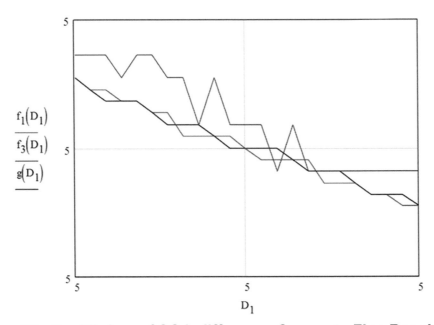

Figure 76. Equilibrium of SG Indifference Curves to Fine Resolution

This shows that the accuracy of the approximate **SG** indifference curve equation $f_3(D_1)$ is as good as that of the true curve $f_1(D_1)$ at the equilibrium point and that both are more accurate than the budget line in this case.

Though the first derivative of equation 75 is not one to one in the domain of D_1 the equilibrium equation once again simplifies to a form that is. The simplified equilibrium equation is therefore necessarily unique. Though we have illustrated this conclusion using our numerical example the analysis is entirely general for all **SG** indifference curves since it depends only on a property of the derivative of equation 75 and not on the parametric conditions.

The form of uniqueness 'proof' or demonstration that we have now seen applied to the **CG**, **SS** and **SG** cases must also apply to any demand function for good 1 even a function that does not depend on the short term demand function of equation 19. Since any such indifference curve for D_2 will be a function, formed by integrating an equation for $- P_1$ and since this expression remains un-simplified, differentiating the indifference function to form an equilibrium equation must result in a multiplicative term equal to $- P_1$ on both sides of the equation by the rule for differentiation of a function of a function.

Since D_2 must always be a one to one function of D_1 in the effective region of any indifference curve because:

$$\frac{d}{dD_1} D_2 < 0$$

for
$$\frac{d}{dD_1} D_2 = \frac{-P_1}{P_2}$$

then the equilibrium solution for any indifference curve for two goods having real (non-discrete) demand must be unique under all possible conditions. Any equilibrium of an indifference curve representing real demand derived from the universally accepted equation 48 must therefore necessarily always be unique.

Though the first derivative of such an indifference function need not be one to one the equilibrium equation would always simplify to a form that is one to one. Note that this property of the first derivative of an indifference function applies to all the first derivatives of all analytical or quasi analytical indifference curve equations used in this work. Furthermore a unique solution would even be obtained for a discontinuous demand function (even though the indifference function is not differentiable) and even at a point of discontinuity as the graph below illustrates.

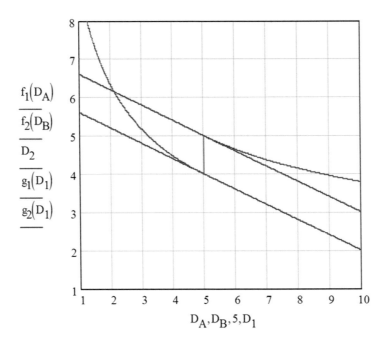

$$f_1(D_A)$$
$$\overline{f_2(D_B)}$$
$$\overline{D_2}$$
$$\overline{g_1(D_1)}$$
$$\overline{g_2(D_1)}$$

$$D_A, D_B, 5, D_1$$

Figure 77. Equilibrium of a Discontinuous Indifference Curve

Figure 77 shows two budget lines in blue at £300 and £350 and the equilibrium conditions for $D_1 = 5$ on the market form of a notional discontinuous indifference curve. Whether or how such an indifference curve might arise need not concern us. Were such a curve to arise we can demonstrate that the equilibrium solution would remain unique even though the indifference curve was discontinuous.

D_A represents demand for good 1 when $D_1 < 5$ and D_B represents demand for good 1 when $D_1 > 5$. Note this makes demand for good 2 indeterminate between 4 and 5 when $D_1 = 5$. Despite this the equilibrium solution is clearly unique. Under the assumptions of ordinal utility theory all points on the indifference curve are of equal utility or value. But, the equilibrium point at $D_1 = 5$, $D_2 = 4$ is clearly preferable to the point at $D_1 = 5$, $D_2 = 5$ or any point in between because it achieves the same level of utility at lower cost. The optimum equilibrium solution is therefore unique once again. As we saw earlier in figure 50 an equivalent formulation could be applied under the assumptions of the new theory of value, though the ordinal utility formulation is doubtless a more familiar approach.

The uniqueness 'proof' or demonstration used here is therefore universal for all real (non-discrete) demand. We will consider the equilibrium between two goods where one or both goods have a discrete demand function when we consider the general equilibrium of real markets (in the economic sense).

Economists have held for over 100 years that the equilibrium solutions need not be unique (except for competitive goods) because the first derivative of the indifference curve equation need not be one to one. Though their analysis is correct their conclusion is not because the equilibrium equation will invariably simplify to a form that is one to one for all goods having real demand.

We have now 'fairly rigorously' demonstrated that the equilibrium solutions, for all the non uni-modal indifference curves derived so far, are necessarily unique. As we shall see the **CG**, **SS** and **SG** cases are the only cases that give rise to non uni-modal indifference curves and therefore the only cases in which the uniqueness of the equilibrium solution is an issue. It follows that the equilibrium solution for all combinations of competitive, speculative and **Giffen** good markets must be unique despite the obvious geometry, shown in figures 49 and 50, which seems to point to the contrary in some cases. The reason why this is so, is that the values of the parameters in all the indifference curve equations constrain the equilibrium point to a unique solution.

This is because the sets of 4 equations with 4 unknowns (the two demand functions for each of the two goods, the indifference curve function and the simplified equilibrium equation) for all possible combinations of competitive, speculative and **Giffen** goods are all determinant systems of simultaneous equations in each of which the values of **P₁**, **P₂**, **D₁** and **D₂** are determined by the parameters and the constants of integration to give unique solutions. Furthermore, the limited available supply of **Giffen** goods, in cases involving such goods, and the resulting constraints on effective demand make unique solutions inevitable.

Hitherto economists were unable to prove a unique equilibrium for speculative and **Giffen** good markets. Now we can. Despite supposed mathematical evidence to the contrary it has always been known that equilibrium solutions must be unique (even though we could not prove it) because real markets (in the economic sense), even speculative and **Giffen** good markets, generally clear, more or less. If equilibrium solutions were not unique for all goods the general equilibrium of real markets would 'always' be unstable because all conceivable real-world general equilibrium conditions must contain examples of such goods. We will return to this topic when we investigate the equilibrium of real markets.

The GG Case

Assuming for the time being that two **Giffen** goods can exist simultaneously in an 'isolated' market then:

$$P_1 = V_1 - C_1 \cdot \ln(D_1) - \frac{V_1}{D_1}$$

$$P_2 = V_2 - C_2 \cdot \ln(D_2) - \frac{V_2}{D_2}$$

Substituting into equation 51:

$$\int V_2 - C_2 \cdot \ln(D_2) - \frac{V_2}{D_2} \, dD_2 = -\int V_1 - C_1 \cdot \ln(D_1) - \frac{V_1}{D_1} \, dD_1 \qquad \text{Equation 76}$$

Therefore:

$$V_2 \cdot D_2 - C_2 \cdot D_2 \cdot \ln(D_2) + C_2 \cdot D_2 - V_2 \cdot \ln(D_2) = -V_1 \cdot D_1 + C_1 \cdot D_1 \cdot \ln(D_1) - C_1 \cdot D_1 + V_1 \cdot \ln(D_1) + k$$

Equation 77

MATHCAD cannot solve this equation analytically but it can be solved, plotted and differentiated iteratively as a numerical function using the MATHCAD root function, in the same way as we did in deriving the true indifference curve in the **SS** and **SG** cases.

The MATHCAD definition of the true numerical form of the **GG** indifference curve is then given by:

$$D_1 = 1, 1.001.. \frac{2V_1}{C_1} \qquad TOL = 10^{-12} \qquad D_2 = \frac{V_2}{C_2}$$

$$f_1(D_1) := root\left[V_2 \cdot D_2 - C_2 \cdot D_2 \cdot \ln(D_2) + C_2 \cdot D_2 - V_2 \cdot \ln(D_2) - \left(-V_1 \cdot D_1 + C_1 \cdot D_1 \cdot \ln(D_1) - C_1 \cdot D_1 + V_1 \cdot \ln(D_1) + k\right), D_2\right]$$

Equation 78

Note that the value of **D₂** defined above is only a starting value or initial estimate of **D₂** used in the iterative solution process. For equilibrium of two **Giffen** goods both traded in bull markets demand must be less than or equal to the competitive equilibrium level of demand. That is:

$$D_1 \leq \frac{V_1}{C_1} \qquad AND \qquad D_2 \leq \frac{V_2}{C_2}$$

We will investigate the equilibrium condition at:

$$D_1 = 4 \qquad D_2 = 3$$

that satisfies these relations. The value of the constant **k** that is appropriate to these conditions in equation 77 is 522.855574171349 and the value of **B** that satisfies these conditions in equation 54 is 377.262976470236. Note that this is effectively the market form of the **GG** indifference curve because it contains only a single constant **k**. The graph below shows the equilibrium condition of the true **GG** indifference curve for the example data.

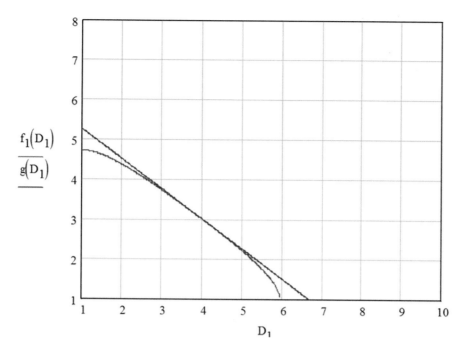

Figure 78. Equilibrium of True Indifference Curve for two Giffen Goods

The true indifference curve for two **Giffen** goods appears to be strictly concave in the domain of D_1. Note however that substitution only occurs over less than two thirds of the domain. $f_1(D_1)$ for our example data in fact becomes unreal when $D_1 > 5.937760060650768$ and $D_2 \approx 1$ confirming the lower bound of the domain of D_2. The lower half of the domain and co-domain corresponds to a **Giffen** good bull market for both goods. At higher levels of demand the **Giffen** good progressively reverts to behaviour similar to a competitive good so figure 79 covers all of the domain that we need to be able to cover. We can confirm that the **GG** indifference curve is strictly concave if we plot the second derivative of $f_1(D_1)$ as shown below:

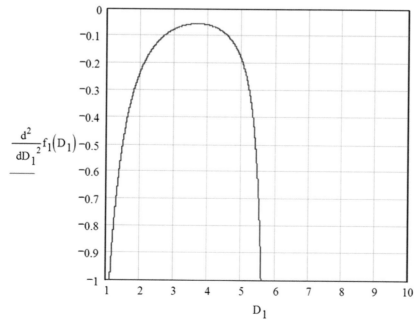

Figure 79. Second Derivative of True GG Indifference Curve

Figure 79 confirms that there is no point of inflection in the true **GG** indifference curve because the second derivative of $f_1(D_1)$ does not intersect the **x** axis at any point in the domain of D_1. The indifference curve for two **Giffen** goods is therefore strictly concave.

The equilibrium equation for the true **GG** indifference curve is given by:

$$\frac{d}{dD_1}f_1(D_1) = \frac{-P_1}{P_2}$$

If we add P_1/P_2 to both sides of this equation and plot the resulting expression against D_1 we can confirm the **GG** equilibrium solution to be unique.

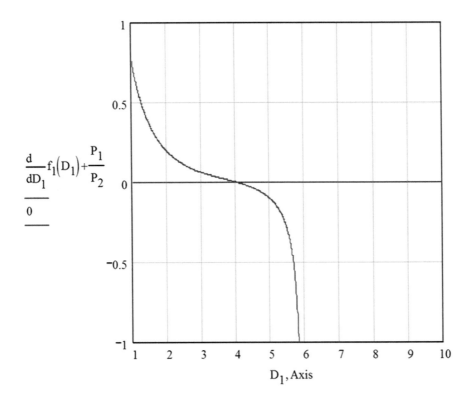

Figure 80. Equilibrium Equation for GG Indifference Curve

MATHCAD can correctly evaluate the root of the equilibrium equation as 4 to 12 decimal places as the table below shows.

$D_1 = $ 3.999999999999   , 3.9999999999991   .. 4   TOL $= 10^{-12}$

$\dfrac{d}{d\,D_1}\, f_1\,(D_1)\;+\;\dfrac{P_1}{P_2}\;=$

4.04121180963557.10^{-14}
2.20934381900406.10^{-14}
7.58282325818982.10^{-14}
8.02691246803988.10^{-14}
-6.99440505513849.10^{-14}
-3.71924713249427.10^{-14}
-5.02931030155196.10^{-14}
3.00870439673417.10^{-14}
-1.94289029309402.10^{-14}
-4.74065231514942.10^{-14}
-1.49880108324396.10^{-14}

Note however that resolution is getting a little ragged here as the fourth row from the bottom of the table is positive not negative. In the **GG** case we have produced an entire indifference curve solution using solely iterative techniques. For completeness it would be helpful if we could also derive a quasi-analytical solution in a similar way to the solutions for the **SS** and **SG** cases even though this is not strictly necessary in the **GG** case.

If we apply an approximation to **Ln(D₂)** in a similar way to that employed in the **SS** and **SG** cases we should once again be able to derive a soluble close approximation to equation 77. The Co-domain of a **Giffen** good market for **D₂** is also the interval (**1, 2V₂/C₂**) because it depends on the demand function for a speculative good. Therefore, the mid-point of the co-domain of **D₂** is also approximately **V₂/C₂**.

Thus, applying the same approximation to **Ln(D₂)** used in earlier cases:

$$V_2 \cdot D_2 - C_2 \cdot D_2 \cdot \left[\ln\left(\frac{V_2}{C_2}\right) + \frac{\left(D_2 - \frac{V_2}{C_2}\right)}{D_2} \right] + C_2 \cdot D_2 - V_2 \cdot \ln(D_2) = -V_1 \cdot D_1 + C_1 \cdot D_1 \cdot \ln(D_1) - C_1 \cdot D_1 + V_1 \cdot \ln(D_1)$$

Equation 79

Solving for **D₂** and applying constants of integration:

$$D_2 = \frac{k_1 \cdot V_2}{\left(-V_2 + C_2 \cdot \ln\left(\frac{V_2}{C_2}\right)\right)} \cdot W\left[\frac{\left(-V_2 + C_2 \cdot \ln\left(\frac{V_2}{C_2}\right)\right)}{V_2} \cdot \exp\left[\frac{\left(V_2 + V_1 \cdot D_1 - C_1 \cdot D_1 \cdot \ln(D_1) + C_1 \cdot D_1 - V_1 \cdot \ln(D_1)\right)}{V_2} \right] \right] + k_2$$

Equation 80

If we ignore the constants of integration for the time being, there are no real values for **D₂** for any value in the domain of **D₁** that satisfy this equation for any positive real values of the parameters because the exponential term is always positive and the term:

$$\left(-V_2 + C_2 \cdot \ln\left(\frac{V_2}{C_2}\right)\right)$$

is always negative. The argument of the **W** function is always less than - 0.367879441171443 and hence the value of the **W** function and thus the value of **D₂** is always unreal or complex. In the absence of the iterative solution, we might have concluded that either two **Giffen** goods cannot exist simultaneously in an 'isolated' market, or that substitution between two such goods cannot take place. Note that a market can be effectively temporarily isolated from the rest of the world market geographically and more permanently politically.

Because **D₂** in equation 80 is unreal we cannot plot equation 80 as a graph. If however we apply a simpler, though unfortunately less rigorous, analytical approximation to equation 77 we should be able to establish why there are no real roots of equation 80. This analytical approximation is still based on a similar idea of substituting **V₂/C₂**, the midpoint of the co-domain of **D₂**, for **D₂**, though in this case for **D₂** outside 'all' logarithmic terms, not inside.

Substituting **V₂/C₂** for **D₂** outside all logarithmic terms of equation 77 gives:

$$\frac{V_2^2}{C_2} - V_2 \cdot \ln(D_2) + V_2 - V_2 \cdot \ln(D_2) = -V_1 \cdot D_1 + C_1 \cdot D_1 \cdot \ln(D_1) - C_1 \cdot D_1 + V_1 \cdot \ln(D_1)$$

Equation 81

Note terms have not been collected to aid comparison with equation 77 and that constants of integration have been omitted.

Solving for **D₂** and applying constants of integration:

$$D_2 = k_3 \cdot \exp\left[\frac{\left(\frac{V_2^2}{C_2} + V_2 + V_1 \cdot D_1 - C_1 \cdot D_1 \cdot \ln(D_1) + C_1 \cdot D_1 - V_1 \cdot \ln(D_1)\right)}{2 \cdot V_2}\right] + k_4$$

Equation 82

If however, we ignore the constants of integration for the time being and plot equation 82 against **D₁** it becomes immediately apparent why there can be no real solution to equation 80.

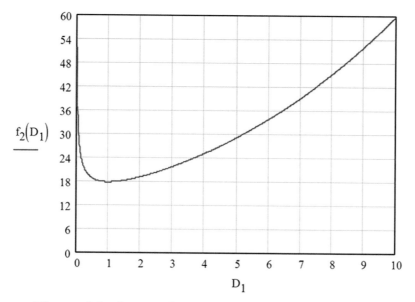

Figure 81. Approximate 'Indifference Curve' for two Giffen Goods

In this case the graph is plotted with a 'true' origin at (0, 0). All that downward sloping part of the curve to the left of **D₁** = 1 is outside the domain of **D₁**. There is therefore no point in the domain of **D₁** for equation 81 that satisfies the condition for equilibrium of equation 48

$$\frac{dD_2}{dD_1} = -\frac{P_1}{P_2}$$

because the slope is positive or zero throughout the domain of **D₁**.

It would be preferable to be able to evaluate constants for equation 80. Unfortunately, however though equation 80 is differentiable, MATHCAD cannot evaluate the constant **k₁** correctly and is unable to converge to a solution for **k₂**.

We are therefore forced to (temporarily) fall back on the simpler and less rigorous simplification of equation 81 used to derive equation 82. If we evaluate the constants of integration of equation 82 in the usual way and re-plot the graph of the equation we find something remarkable has happened. Differentiating equation 82 gives:

$$\frac{d}{dD_1}D_2 = \frac{1}{2} \cdot k_3 \cdot \frac{\left(V_1 - C_1 \cdot \ln(D_1) - \frac{V_1}{D_1}\right)}{V_2} \cdot \exp\left[\frac{1}{2} \cdot \frac{\left(\frac{V_2^2}{C_2} + V_2 + V_1 \cdot D_1 - C_1 \cdot D_1 \cdot \ln(D_1) + C_1 \cdot D_1 - V_1 \cdot \ln(D_1)\right)}{V_2}\right]$$

This expression must be equated to – **P₁/P₂** and solved for **k₃** and thence **k₄** evaluated from equation 82 for the equilibrium values of **D₁** and **D₂** to obtain the particular solution required for the **GG** case. The   values   of   **k₃**   and

k₄ appropriate to this equilibrium condition are - 0.20380258355440792922 and 8.10186388447261 respectively.

The remarkable effect of applying these constants to equation 82 is to 'invert' (in the non-mathematical sense - meaning turn upside down) the graph of figure 82 as well as correcting the slope and repositioning the curve. This is not the first time that an 'inverted' function in this sense has arisen when considering **Giffen** goods. The graph below shows the resulting equilibrium from equation 82.

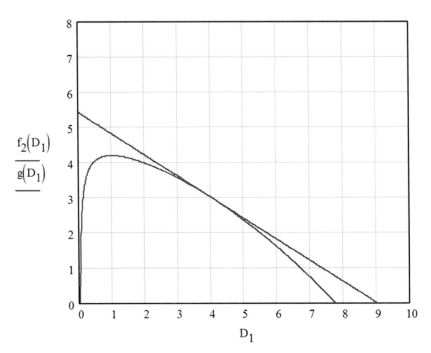

Figure 82. Equilibrium of Approximate GG Indifference Curve

This graph is also drawn using a 'true' origin at (0, 0) to illustrate the inversion effect. Indifference curves of this form though unusual are not entirely unknown. Figure 82 is reminiscent of the indifference curve implied by **Becker's** theory concerning an altruistic head of household (**Garry Becker**, 1991) though the graph is less symmetrical about the 'maximum' and the axes have been arbitrarily reversed.

It would seem that with this case we have reached the analytical limit of use of **Lambert's W** function in MATHCAD, which is unable to perform this 'inversion' manoeuvre for equation 80. We can however obtain a result similar to figure 82 if we 'help' 'inversion' by inserting a minus sign prior to terms both inside and outside the **W** function. Note that the term immediately outside the **W** function is the reciprocal of the first term inside the **W** function. This form of 'symmetry' applies to all the **W** functions used in this paper. Inserting a minus sign both inside and outside the **W** function maintains this 'symmetry'. This is therefore rather like inserting a constant – **k₁** instead of **k₁** into an equation that did not involve a **W** function. Use of a constant – **k₁** instead of **k₁** would of course have no effect on the final form of an equation once constants were evaluated. Equation 80 then takes the form shown below:

$$D_2 = \frac{k_1 \cdot V_2}{\left(V_2 - C_2 \cdot \ln\left(\frac{V_2}{C_2}\right)\right)} \cdot W\left[\frac{\left(V_2 - C_2 \cdot \ln\left(\frac{V_2}{C_2}\right)\right)}{V_2} \cdot \exp\left[\frac{\left(V_2 + V_1 \cdot D_1 - C_1 \cdot D_1 \cdot \ln(D_1) + C_1 \cdot D_1 - V_1 \cdot \ln(D_1)\right)}{V_2}\right]\right] + k_2$$

Equation 83

Differentiating equation 83 yields:

$$\frac{d}{dD_1}D_2 =$$

$$\frac{-k_1 \cdot \left(V_1 - C_1 \cdot \ln(D_1) - \frac{V_1}{D_1}\right)}{\left(-V_2 + C_2 \ln\left(\frac{V_2}{C_2}\right)\right)} \cdot \frac{W\left[\frac{-1}{V_2}\left(-V_2 + C_2 \cdot \ln\left(\frac{V_2}{C_2}\right)\right) \cdot \exp\left[\frac{\left(V_2 + V_1 \cdot D_1 - C_1 \cdot D_1 \cdot \ln(D_1) + C_1 \cdot D_1 - V_1 \cdot \ln(D_1)\right)}{V_2}\right]\right]}{\left[1 + W\left[\frac{-1}{V_2}\left(-V_2 + C_2 \cdot \ln\left(\frac{V_2}{C_2}\right)\right) \cdot \exp\left[\frac{\left(V_2 + V_1 \cdot D_1 - C_1 \cdot D_1 \cdot \ln(D_1) + C_1 \cdot D_1 - V_1 \cdot \ln(D_1)\right)}{V_2}\right]\right]\right]}$$

This expression must be equated to − P_1/P_2 and solved for k_1 and thence k_2 evaluated from equation 83 for the equilibrium values of D_1 and D_2 to obtain the particular solution required for the **GG** case. For the equilibrium condition given above the appropriate values of k_1 and k_2 are - 2.73109068408927 and 9.54626566071262 respectively.

If we plot the equilibrium of both equations 82 and 83 together, we obtain the result shown below.

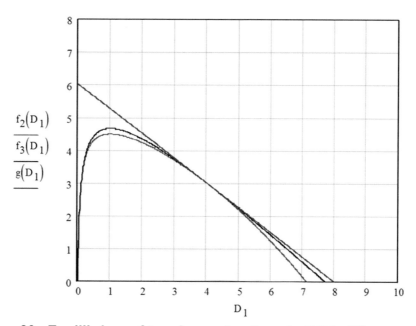

Figure 83. Equilibrium of two Approximations to GG Indifference Curves

$f_3(D_1)$ derived from equation 82 plotted in black is a close quasi-analytical approximation to the true indifference curve and $f_2(D_1)$ derived from equation 81 plotted in brown is a crude analytical approximation. Note the similarity of the two curves. This seems to vindicate our rather irregular analytical procedure used to derive equation 82 and the simpler analytical approximation used to derive equation 81. If we now compare the equilibrium of the true indifference curve $f_1(D_1)$ with $f_3(D_1)$ plotted over the predicted domain of D_1 and co-domain of D_2 we obtain the result shown below.

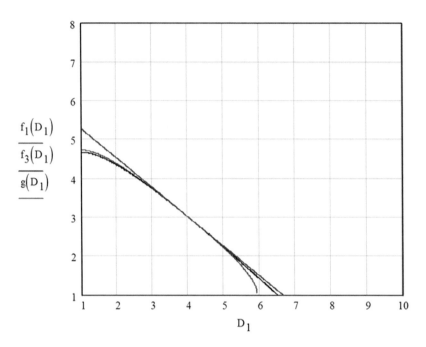

Figure 84. Equilibrium of Approximate & True GG Indifference Curve

Because of the complexity (in the non-mathematical sense) of the **W** function, to produce this graph across the whole plain the tolerance, **TOL** in the **W** function definition, must be increased from 10^{-15} to 10^{-12}. The graphs peter out part way across the plane if you do not, due to a convergence problem. Manipulating other parameters in the definition of **Lambert's W** function seems to have little or no effect on the outcome for any of the indifference curves considered in this work.

The accuracy of the equilibrium solution is shown in figure 85 below, which represents figure 84 plotted to 14 decimal place resolution.

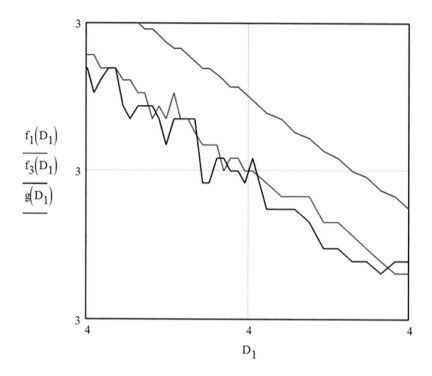

Figure 85. Indifference Curves for the GG Case to Fine Resolution

At equilibrium both indifference curves are of similar accuracy and both are more accurate than the budget line.

Inserting an error correcting factor **E** and applying the correction procedure to equation 83 that we used to improve the quasi-analytical approximations in the **SS** and **SG** cases seems not to improve the accuracy of approximation in the **GG** case. No amount of 'pulling' the approximate indifference curve with increasing or reducing values of **E** seems to be able to correct the deviation of $f_3(D_1)$ from $f_1(D_1)$ for values of D_1 above the mid-point of the domain of a **Giffen** good. In some respects, this seems to make matters worse when we consider the first and second derivatives of $f_3(D_1)$.

Readers may recall that at the mid-point of the domain of demand of a **Giffen** good its behaviour begins to revert to a form similar to a competitive good. When demand is below V_1/C_1 a bull market will exist for good **1** and it is this aspect of the behaviour of a **Giffen** good that is of most interest. In this region equation 83 achieves similar accuracy to that achieved in other cases without the need for error correction. In our example over the domain of demand of a **Giffen** good bull market the maximum error is < 1.3% to 1 decimal place.

At $D_1 > 5.937760060650768$ the error becomes infinite (is undefined). It would seem therefore that we should restrict the applicability of equation 83 to the domain of demand of a **Giffen** good bull market (**1**, V_1/C_1). Examining graphs of the first and second derivatives of $f_3(D_1)$ confirms this.

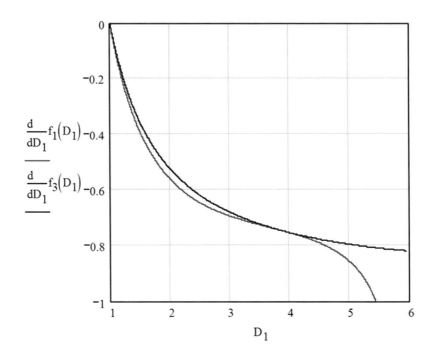

Figure 86. First Derivative of True & Approximate GG Indifference Curves

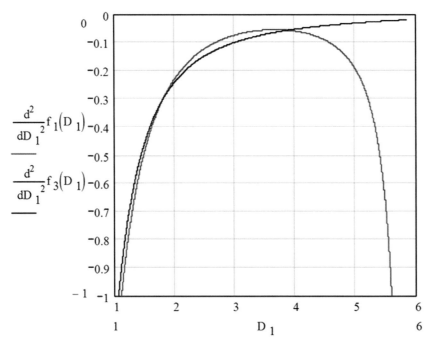

Figure 87. Second Derivative of True & Approximate GG Indifference Curve

If an equilibrium is to be investigated at values of D_1 above this level [any point in the interval [$D_1 \ \varepsilon \ (4, 5.937760060650768)$] the simplest way to accomplish this is to exchange the axes plotting D_1 on the **y** axis and D_2 on the **x** axis. This can be achieved by substituting suffix **1** for suffix **2** and vice versa in equation 83 and re-evaluating the constants k_1 and k_2. We do not need to re-evaluate the constant **k** because **k** is a true constant of integration.

The appropriate values of k_1 and k_2 for our example data are: — 1.98921931711926922124 and 11.5845265897653 respectively. The figure bellow shows the resulting graph.

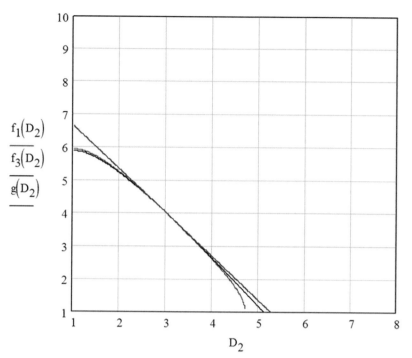

Figure 88. Exchange of Axes of Figure 84

Where:

$$g(D_2) = \frac{B}{P_1} - \frac{P_2}{P_1} \cdot D_2$$

The accurate region of the curve of figure 88 spans the bull market for good **2** (**D₂** < 3) whereas the accurate region of figure 84 spans the bull market for good **1** (**D₁** < 4). Taking the two graphs together the predicted domain and co-domain of **D₁** and **D₂** is accurately spanned.

Equations 82 and 83 and figures 78, 81, 82, 83, 84, 85 and 88 show that it is possible to achieve equilibrium between two **Giffen** goods confirming our initial assumption. Such an occurrence remains an extremely rare event however. The emergence of even a single **Giffen** good is rare. **Giffen** goods generally only arise for staple foodstuffs among the very poor in times of famine. The probability of two **Giffen** goods arising in an isolated market at the same time will be very low indeed. To a first approximation this would be the compound probability or square of probability of a single **Giffen** good arising (a very small number) if we discount the effect that weather conditions or pests that cause one staple crop to fail may also cause failure of another. Though this dual failure seems quite possible it is not in fact culturally very likely. Each of the main groups of staple foods (maize, rice, grain and potatoes) tends to be associated with particular regions of the world and particular populations. The result for the all **Giffen** good case enables us to complete a logical sequence of results for all 6 possible cases of interaction between competitive, speculative and **Giffen** goods.

Summary of Partial Equilibrium Results

We have used examples (though not just those included here) to demonstrate many of our conclusions and so cannot claim the results that depend upon examples as definitive. However, all of the derivation and analysis both economic and mathematical has been general and reasonably rigorous. Much of the theory and particularly that relating to the uniqueness of equilibrium solutions could arguably be held to constitute a proof. Our main conclusions are:

1. For a combination of two competitive goods the indifference curve is strictly convex.
2. For a combination of a competitive good and a speculative good the indifference curve is convex but intersects the x axis.
3. For a combination of a competitive or speculative good and a **Giffen** good the indifference curve is di-modal and intersects the x axis.
4. For a combination of two speculative goods the indifference curve is tri-modal.
5. For a combination of two **Giffen** goods the indifference curve is strictly concave.
6. For all possible cases the equilibrium solution is necessarily unique.

As this analysis has been very long and difficult readers may have lost sight of our objective of defining indifference curves for all possible combinations of competitive, speculative and **Giffen** goods. It may also be unclear which results achieved that objective. We will therefore summarise the main results. The most accurate analytical or quasi-analytical equations for each of the 6 cases of indifference curves for an individual are listed below:

For the **CC** case

$$D_2 = k_1 \cdot \exp\left[\frac{\left(-V_1 \cdot \ln(D_1)\right)}{V_2}\right] + k_2 \qquad\qquad \text{Equation 55}$$

For the **CS** case

$$D_2 = k_1 \cdot \exp\left[\frac{\left(-C_1 \cdot D_1 \cdot \ln(D_1) + C_1 \cdot D_1 - V_1 \cdot \ln(D_1)\right)}{V_2}\right] + k_2 \qquad\qquad \text{Equation 59}$$

For the **CG** case

$$D_2 = k_1 \cdot \exp\left[\frac{\left(-V_1 \cdot D_1 + C_1 \cdot D_1 \cdot \ln(D_1) - C_1 \cdot D_1 + V_1 \cdot \ln(D_1)\right)}{V_2}\right] + k_2 \qquad\qquad \text{Equation 64}$$

All of these results are exact analytical solutions.

'Exact' indifference curves for cases where neither good is competitive can be calculated and plotted iteratively. A reasonably accurate quasi-analytical approximation to these 'true' indifference curves 'loosely' based on **Taylor's** expansion can be achieved using **Lambert's W** function in the following equations. Each of these equations can be used to demonstrate the uniqueness of solutions in every case.

For the **SS** case

$$D_2 = \frac{k_1}{\frac{C_2}{V_2} \cdot \ln\left(\frac{E \cdot V_2}{C_2}\right)} \cdot W\left[\frac{C_2}{V_2} \cdot \ln\left(\frac{E \cdot V_2}{C_2}\right) \cdot \exp\left[\frac{\left(E \cdot V_2 - C_1 \cdot D_1 \cdot \ln(D_1) + C_1 \cdot D_1 - V_1 \cdot \ln(D_1)\right)}{V_2}\right]\right] + k_2$$

Equation 69

For the **SG** case

$$D_2 = \frac{k_3}{\frac{C_2}{V_2} \cdot \ln\left(\frac{E \cdot V_2}{C_2}\right)} \cdot W\left[\frac{C_2}{V_2} \cdot \ln\left(\frac{E \cdot V_2}{C_2}\right) \cdot \exp\left[\frac{\left(E \cdot V_2 - V_1 \cdot D_1 + C_1 \cdot D_1 \cdot \ln(D_1) - C_1 \cdot D_1 + V_1 \cdot \ln(D_1)\right)}{V_2}\right]\right] + k_4$$

Equation 75

For the **GG** case

$$D_2 = \frac{k_1 \cdot V_2}{\left(V_2 - C_2 \cdot \ln\left(\frac{V_2}{C_2}\right)\right)} \cdot W\left[\frac{\left(V_2 - C_2 \cdot \ln\left(\frac{V_2}{C_2}\right)\right)}{V_2} \cdot \exp\left[\frac{\left(V_2 + V_1 \cdot D_1 - C_1 \cdot D_1 \cdot \ln(D_1) + C_1 \cdot D_1 - V_1 \cdot \ln(D_1)\right)}{V_2}\right]\right] + k_2$$

Equation 83

The maximum error (for our example data at least) in these 3 equations lies between 1% and 2.6 % (or < 1.7% when **E** is 'optimised' for predictive accuracy) over the predicted domain of the **SS** and **SG** curves and without correction over the domain of a **Giffen** good bull market for the **GG** curve.

Note that for each successive case the equations become more complicated and the signs of some similar terms are reversed. Equations 55, 59 and 64 can be evaluated and plotted using quite a slow computer and spreadsheet software and can even be evaluated using a scientific calculator. Equations 69, 75 and 83 require a fairly fast computer and mathematics software capable of evaluating and plotting **Lambert's W** function.

As predicted in all 6 cases this approach enables us to plot and calculate equilibrium solutions to very great accuracy though in more difficult cases it has been necessary to use close quasi-analytical approximations to demonstrate the uniqueness of equilibrium solutions.

Some may claim that this set is incomplete because it does not include goods traded in monopoly or oligopoly markets. These markets are not analysed separately in this context because they both depend on the same demand function as goods traded in competitive markets. It is the demand functions of the two goods involved in a partial equilibrium that determines the resulting form of their indifference curves. When we consider general equilibrium however, we will include monopoly and oligopoly markets separately as sub-sets of the competitive equilibrium because what determines that outcome is how prices are determined not just the demand function.

Chapter 11. General Equilibrium & the New Theory of Value

(**Leon Walras**, 1874 & 7) derived the conditions for general equilibrium or clearance of all markets in a perfectly competitive economy. His approach involved a system of n − 1 simultaneous functions of the form:

$$Z_i = f(U, R, T, P_1, P_2, \ldots\ldots, P_{n-1}) \qquad \text{for each good i}$$

Where	Z_i is the excess demand (positive or negative) for good i
	U is preference
	R is initial endowment
	T is technology
And	$P_1, P_2, \ldots\ldots, P_{n-1}$ are n -1 relative prices expressed with respect to the price P_n of the numeraire good n.

Walras concluded that as there were **n - 1** equations with **n - 1** unknown relative prices the system was determinant. His solution was dependent on **Walras' Law** that states that the value of the sum of all excess demands must equal zero and a price adjustment rule that prices should be raised if there was positive excess demand and reduced if there was negative excess demand.

He was unable to propose a practically believable process by which equilibrium was achieved in the market. He suggested a process of 'tatonnement' by which prices were adjusted, modelled by a notional **Walrasian** auctioneer. In his system there was no borrowing and no money.

As we have now established that the partial equilibrium of any pair of goods with real (as opposed to discrete) demand is unique we no longer need to restrict our analysis to competitive markets. This analysis therefore includes all competitive, speculative and **Giffen** goods with real demand traded in any market irrespective of how prices are determined in that market. As we have seen, this holds for goods with discontinuous indifference curves provided only that all goods have real demand. The analysis will even include goods traded under monopoly or oligopoly conditions because, by the definition of a competitive market used in this work, monopoly and oligopoly markets are just special cases of competitive markets because they depend on the competitive demand function. This contention is demonstrated later. We will postpone consideration of goods having discrete demand until we review the general equilibrium of 'real' markets after this analysis is complete.

Note in economics real markets do not have real demand, that is continuous demand, expressed in real numbers, but incremental discrete demand because all goods are traded either in whole numbers of goods or statutory units of measure.

Since this work is solely concerned with value and demand and not supply side economics the system of equations developed so far does not include the markets for the factors of production: capital, labour and land or other material or service inputs to production. These markets can however be simply encompassed in the system by extending the set of goods and services considered to include all goods and services traded in the economy.

We do not need to consider the markets for factors of production as homogeneous either. We can include as many differentiable categories of each as is necessary. The value of **V** is now extended to represents the value of all expenditure on all goods and services with real demand traded in the economy and not just the value of all consumption.

While we have considered how consumers can 'optimise' the allocation of their resources to competing uses, the new theory of value and demand has not defined how a firm could 'optimise' the allocation of its resources. Obviously, a compatible theory of production and the firm could be developed. For the time being however we will circumvent this difficulty by the simple device of assuming that firms have some criteria of preference among their purchases and some basis of 'optimising' the value of goods and services they buy.

If, for example, we assume a firm values its purchases in proportion to their productivity then the system already developed can be simply extended. The new theory of value and demand will then work just as well for a firm as for a consumer. This formulation is not dependent on the assumption of any particular form of function governing a firm's acquisitions, or on the assumption that their needs are satiable, only that whatever function does apply to them includes their preferences and priorities. Neither is it dependent on any particular assumptions concerning returns to scale or the shape of a firm's cost function. Furthermore as we have shown that equilibrium solutions in all markets having real demand are necessarily always unique, the resulting solution is unaffected by non-convexities.

Under the new theory of value we can derive the conditions for general equilibrium using the methods applied in deriving equation 44. Because we seek the conditions for general equilibrium rather than just a partial equilibrium, this requires we relax (at least initially) the usual simplifying assumption of the new theory of value and demand concerning no substitution between goods that are not close substitutes.

now
$$V = \sum_{x=1}^{n} P_x D_x$$
Equation 84

where
V is the total value of all transactions in the economy
P_x is the price of good **x**
D_x is demand for good **x**
n is the number of goods and services in the market

and
x is an arbitrary good

A complication of this relationship often arises in real markets (in the usual sense) because of the common business practice of offering quantity discounts, lower prices for purchases of higher volumes, specially negotiated prices for very large orders or to firms conducting a large volume of other business with a supplier. Where this occurs the volume of each good sold at each price must be treated as a separate good for equation 84 to hold.

Due to the circular flow of income:

$$V = Y$$

where $\quad\quad\quad\quad\quad Y$ is National Income

therefore $\quad\quad\quad\quad\quad Y = \sum_{x=1}^{n} P_x D_x$ $\quad\quad\quad\quad$ Equation 85

As we have seen equations 84 and 85 now reflect both consumers' and by analogy now all buyers' assets or initial endowments, income, borrowing, preferences, priorities and 'utility' directly. Deriving the equilibrium conditions under the new theory of values is therefore an exact parallel with **Walras'** use of the variables **U** and **R**. Furthermore since the behaviour of firms could be assumed to be dependent on the productivity of their resources such behaviour would reflect their technology **T**, a major determinant of productivity. Finally the equilibrium system of the new theory of value and demand does not require any notion of excess demand or depend on relative prices, since as we shall see it includes demand, prices, for all goods and services, incomes and money directly. It therefore includes the markets for domestic and commercial borrowing and other sources of finance for industry and Government. We have already shown that borrowing can be treated as just another good. Money can also be included in the system as a good by expressing the price of money as **1** (or −**1** if you prefer because money flows in the opposite direction to goods, either will work). This follows from the fact that all prices are expressed in terms of money in the form of the national currency within each nation's economy. $D_£$ and $V_£$ (both are actually the same) then represent the total demand for, and 'consumption' of, money in the UK economy.

Differentiating equation 84 by demand D_r for each good **r** gives a system of **n** simultaneous differential equations (each containing **2n +1** terms) of the form:

$$\frac{dV}{dD_r} = \sum_{x=1}^{r-1}\left(P_x\frac{dD_x}{dD_r} + D_x\frac{dP_x}{dD_r}\right) + D_r\frac{dP_r}{dD_r} + P_r + \sum_{x=r+1}^{n}\left(P_x\frac{dD_x}{dD_r} + D_x\frac{dP_x}{dD_r}\right)$$

$\quad\quad\quad\quad\quad\quad\quad\quad\quad\quad\quad\quad\quad\quad\quad\quad\quad\quad\quad\quad$ Equations 86

This system of equations defines the behaviour of consumers and other buyers substitution between goods and services to maximise their utility or value since it contains **2.$^n C_2$** terms of the form $\underline{\frac{dD_x}{dD_r}}$ the marginal rate of substitution, or for a firm, technical substitution, between each pair of goods resulting from a change in the quantity of good **r** consumed or used. Equations 86 are therefore the set of unconstrained equilibrium equations we seek to optimise. These equations are necessarily linearly independent because they are defined to be unconstrained and because they each contain a balancing term of the form:

$$\frac{dV}{dD_r}$$

If we assume that a set of constraints applies to equations 86 reflecting that the total value of all consumption and use of all goods does not change in the short term, and the value of increased consumption or use of any good **r** is equal and opposite to the total decrease in the net value of consumption and use of all other goods, we can simplify equations 86 as shown below. Such constraints can be represented by the relationship given in equations 87 for each good **r** as a system of **n** simultaneous differential equations (each containing **n** terms) of the form:

$$D_r\frac{dP_r}{dD_r} + \sum_{x=1}^{r-1} D_x\frac{dP_x}{dD_r} + \sum_{x=r+1}^{n} D_x\frac{dP_x}{dD_r} = 0 \qquad \text{Equations 87}$$

that is shown to be correct hereafter. The system of equations 87 are necessarily linearly dependent because each such equation is a constraint and because each of these equations contains no balancing term.

Equations 87 are analogous to **Walras' Law** and could also be stated as: the total net excess 'expenditure or consumption' and use (or value) in the economy must always be equal to zero. We will also consider how a system of equations similar to equations 86 can be formed by differentiating equation 84 by each price **Pr** instead of the level of demand **Dr**. In that formulation the equivalent relationship to equations 87 which defines excess value of supply is given by the sum of the products of the equilibrium price of each good and the first derivative of demand for each good with respect to the price **Pr**. This formulation is exactly the same as the **Walrasian** product $P_i.P_n.Z_i$ for each good **i**.

The price of each good **r** is then given by a system of **n** simultaneous differential equations (each containing **n +1** terms) of the form:

$$\frac{dV}{dD_r} = P_r + \sum_{x=1}^{r-1} P_x\frac{dD_x}{dD_r} + \sum_{x=r+1}^{n} P_x\frac{dD_x}{dD_r} \qquad \text{Equations 88}$$

Equations 88 must therefore be a set of linearly independent equations because equations 86 were defined to be linearly independent and because equations 88 contain the same balancing terms as equations 86.

The assumptions concerning the form of equations 87 and 88 can be validated by integrating equation 88 as follows:

$$\int \frac{dV}{dD_r}, dD_r = \int P_r, dD_r + \sum_{x=1}^{r-1} \int P_x\frac{dD_x}{dD_r}, dD_r + \sum_{x=r+1}^{n} \int P_x\frac{dD_x}{dD_r}, dD_r$$

Equation 89

Integrating by parts:

$$\int P_x\frac{dD_x}{dD_r}, dP_r = P_xD_x - \int D_x\frac{dP_x}{dD_r}, dD_r$$

Equation 90

Substituting into equation 88

$$V = \int P_r, dD_r + \sum_{1}^{n}P_xD_x - P_rD_r - \sum_{x=1}^{r-1}\int D_x\frac{dP_x}{dD_r}, dD_r - \sum_{x=r+1}^{n}\int D_x\frac{dP_x}{dD_r}, dD_r$$

but from equation 87

$$D_r\frac{dP_r}{dD_r} = -\sum_{x=1}^{r-1}D_x\frac{dP_x}{dD_r} - \sum_{x=r+1}^{n}D_x\frac{dP_x}{dD_r}$$

therefore
$$V = \int P_r, dD_r + \sum_{x=1}^{n}P_xD_x - P_rD_r + \int D_r, dP_r$$

now
$$\int P_r, dP_r = P_rD_r - \int D_r, dD_r$$

therefore
$$V = \sum_{x=1}^{n}P_xD_x \qquad \text{is TRUE}$$

Thus equations 87 are correct since this is the form of equations 84 the functions from which they were derived. Equations 87 is then given by equations 86 minus equations 88.

Provided that
$$\frac{dV}{dD_r} \neq 0$$

equations 88 are, as we shall see, a determinant system of **n** linearly independent differential equations expressing the price of all goods in the economy in terms of functions of the price of all other goods, the marginal rate of substitution between each of $^nC_2$ pairs of goods and the marginal rate of change in the value or utility of all purchases of all **n** goods in the economy with respect to changes in the demand for each good.

One immediate consequence of this is that **V** though constrained by equations 87 cannot be constant in equations 88. This must be so because otherwise the general equilibrium of the whole economy would be linearly dependent, singular and therefore 'effectively not calculable' and markets would not clear. Since it is evident that real markets (in the economic sense) do clear, more or less, then the representative form of equation 88 must be linearly independent and therefore non-singular. Clearly National Income and the National Product is not constant but is always changing and can be expected to be influenced both in real and money terms by changes in the prices of all goods and changes in the real demand for and output of all goods. The behaviour of

$$\frac{dV}{dD_r}$$

in this system will nevertheless be somewhat unexpected.

The implication of the linear independence of equations 88 is that the price of every good in the economy must be independent of the price of every other good. We will demonstrate that this conclusion is correct below using a simulation of the equilibrium solution of equations 88.

This rather technical, mathematical argument and the resulting conclusion has the surprising implication that no substitution can take place between (in principle) any pair of goods in a general equilibrium. We will show this to be so analytically and when using the same simulation of the general equilibrium process.

The system of equations 88 is independent of the form of the demand function for any good and therefore defines not just general equilibrium in a perfectly competitive economy but general equilibrium in any economy that contains only goods with real demand. The analysis is not therefore dependent on the form of short-term demand function or the demand functions for speculative or **Giffen** goods derived elsewhere in this work.

To help us to solve equations 88 we can represent this system in matrix form as:

$$\underline{V} = D.\underline{P} \qquad\qquad \text{Equation 91}$$

where
$$\underline{V} = \begin{pmatrix} \dfrac{d}{dD_1}V \\[2ex] \dfrac{d}{dD_2}V \\[2ex] \dfrac{d}{dD_3}V \\[1ex] \cdot \\ \cdot \\ \cdot \\ \cdot \\ \cdot \\ \cdot \\[1ex] \dfrac{d}{dD_n}V \end{pmatrix}$$
is a vector of differential coefficients

$$D = \begin{pmatrix} 1 & \dfrac{d}{dD_1}D_2 & \dfrac{d}{dD_1}D_3 & \cdots & \cdot & \dfrac{d}{dD_1}D_{n-1} & \dfrac{d}{dD_1}D_n \\[2ex] \dfrac{d}{dD_2}D_1 & 1 & \dfrac{d}{dD_2}D_3 & \cdots & \cdot & \dfrac{d}{dD_2}D_{n-1} & \dfrac{d}{dD_2}D_n \\[2ex] \dfrac{d}{dD_3}D_1 & \dfrac{d}{dD_3}D_2 & 1 & \cdots & \cdot & \dfrac{d}{dD_3}D_{n-1} & \dfrac{d}{dD_3}D_n \\[1ex] \cdot & \cdot & \cdot & \cdots & \cdot & \cdot & \cdot \\ \cdot & \cdot & \cdot & \cdots & \cdot & \cdot & \cdot \\ \cdot & \cdot & \cdot & \cdots & \cdot & \cdot & \cdot \\ \cdot & \cdot & \cdot & \cdots & \cdot & \cdot & \cdot \\[1ex] \dfrac{d}{dD_n}D_1 & \dfrac{d}{dD_n}D_2 & \dfrac{d}{dD_n}D_3 & \cdots & \cdot & \dfrac{d}{dD_n}D_{n-1} & 1 \end{pmatrix}$$

is a matrix of differential coefficients

and $\quad\underline{P} =$ 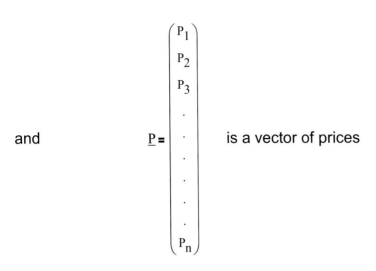 is a vector of prices

for all **n** goods and services in the economy.

Then $\qquad \underline{P} = D^{-1}.\underline{V} \qquad\qquad$ Equation 92

gives the general equilibrium solution for all prices of all goods in the economy. MATHCAD cannot invert the matrix **D** in the general form given above but because all the coefficients in the matrix are differential coefficients and not partial differential coefficients and each term in equation 84 contains only the price and demand of a single good we can substitute all terms:

$$\frac{d}{dD_r}D_x$$

by $\qquad\qquad \dfrac{-P_r}{P_x} \qquad\qquad\qquad$ from equation 48

The matrix **D** is then given by:

$$
D = \begin{pmatrix}
1 & \dfrac{-P_1}{P_2} & \dfrac{-P_1}{P_3} & \cdots\cdots & \dfrac{-P_1}{P_{n-1}} & \dfrac{-P_1}{P_n} \\[2ex]
\dfrac{-P_2}{P_1} & 1 & \dfrac{-P_2}{P_3} & \cdots\cdots & \dfrac{-P_2}{P_{n-1}} & \dfrac{-P_2}{P_n} \\[2ex]
\dfrac{-P_3}{P_1} & \dfrac{-P_3}{P_2} & 1 & \cdots\cdots & \dfrac{-P_3}{P_{n-1}} & \dfrac{-P_3}{P_n} \\[2ex]
\cdot & \cdot & \cdot & \cdots\cdots & \cdot & \cdot \\
\cdot & \cdot & \cdot & \cdots\cdots & \cdot & \cdot \\
\cdot & \cdot & \cdot & \cdots\cdots & \cdot & \cdot \\
\cdot & \cdot & \cdot & \cdots\cdots & \cdot & \cdot \\
\cdot & \cdot & \cdot & \cdots\cdots & \cdot & \cdot \\
\dfrac{-P_n}{P_1} & \dfrac{-P_n}{P_2} & \dfrac{-P_n}{P_3} & \cdots\cdots & \dfrac{-P_n}{P_{n-1}} & 1
\end{pmatrix}
$$

This in no way constrains the solution because the slope of the market form of all true indifference curves is equal to the ratio – P_r/P_x at every point in the domain of D_r, as we demonstrated in the last chapter for all forms of demand function derived in this work. The solution obtained here is not therefore a special case of general equilibrium but an entirely general solution that applies under all circumstances. We can simply demonstrate this contention for each of the 6 indifference curves derived earlier by plotting the marginal rate of substitution and the ratio – P_r/P_x together over the domain of D_r. To do this we need to define the ratio – P_r/P_x in terms of D_r.

The marginal rate of substitution is defined as the first derivative of $f_1(D_1)$, the market form of the true indifference curve for each of the 6 cases given above. The ratio – P_1/P_2 is then defined as $f_2(D_1)$ as follows:

$$
f_2(D_1) = \dfrac{-\dfrac{V_1}{D_1}}{\dfrac{V_2}{f_1(D_1)}} \qquad \text{in the \textbf{CC} case}
$$

When $f_2(D_1)$ and the first derivative of $f_1(D_1)$ for any of the 6 cases are plotted together the two curves are identical to such accuracy that only the curve plotted second is visible. This even applies to purely numerical functions. Since MATHCAD can fail to plot a curve altogether under some circumstances, to confirm that both curves have indeed been plotted we need to force them apart. This has been achieved in all 6 cases by adding a small increment to D_1 in the **x** direction and by subtracting a small increment from $f_2(D_1)$ in the **y** direction. The resulting graph in the **CC** case is shown below:

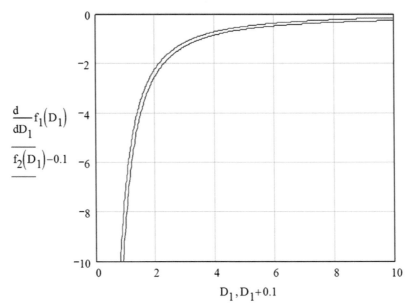

Figure 89. Slope of the CC Indifference Curve and Ratio – P₁/P₂

The ratio – **P₁/P₂** is defined as:

$$f_2(D_1) = \frac{-\left(C_1 \cdot \ln(D_1) + \dfrac{V_1}{D_1}\right)}{\dfrac{V_2}{f_1(D_1)}}$$

in the **CS** case

and the resulting graph is given below:

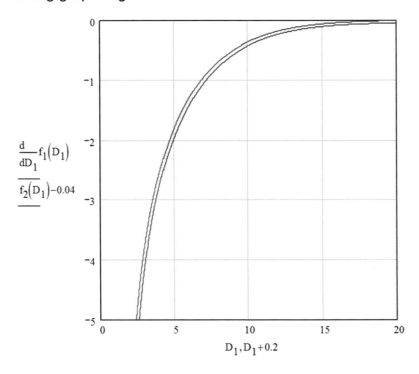

Figure 90. Slope of the CS Indifference Curve and Ratio – P₁/P₂

The ratio – **P₁/P₂** is defined as:

$$f_2(D_1) = \frac{-\left(V_1 - C_1 \cdot \ln(D_1) - \dfrac{V_1}{D_1}\right)}{\dfrac{V_2}{f_1(D_1)}}$$

in the **CG** case

and the resulting graph is given below.

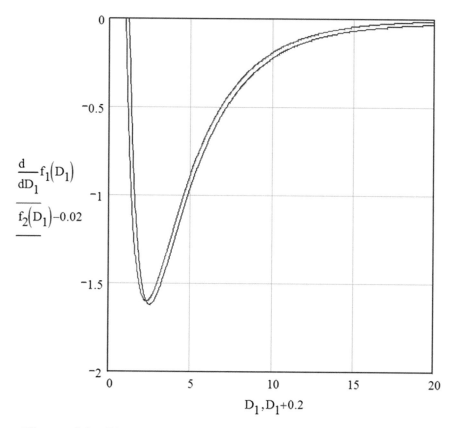

Figure 91. Slope of the CG Indifference Curve and Ratio – P₁/P₂

The ratio – **P₁/P₂** is defined as:

$$f_2(D_1) = \frac{-\left(C_1 \cdot \ln(D_1) + \dfrac{V_1}{D_1}\right)}{C_2 \cdot \ln(f_1(D_1)) + \dfrac{V_2}{f_1(D_1)}}$$

in the **SS** case

and the resulting graph is given below.

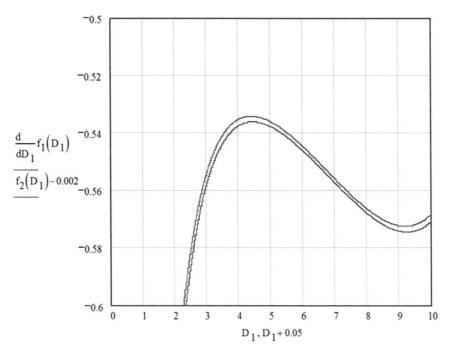

Figure 92. Slope of the SS Indifference Curve and Ratio – P₁/P₂

The ratio – **P₁/P₂** is defined as:

$$f_2(D_1) = \frac{-\left(V_1 - C_1 \cdot \ln(D_1) - \dfrac{V_1}{D_1}\right)}{C_2 \cdot \ln(f_1(D_1)) + \dfrac{V_2}{f_1(D_1)}} \qquad \text{in the \textbf{SG} case}$$

and the resulting graph is given below.

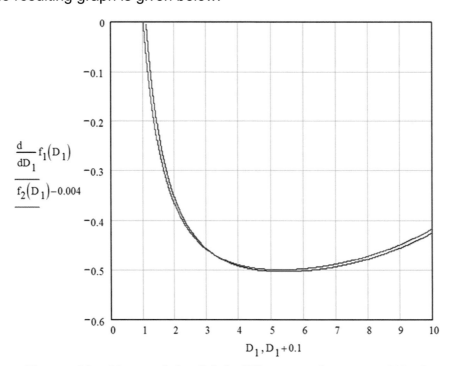

Figure 93. Slope of the SG Indifference Curve and Ratio – P₁/P₂

The ratio $- P_1/P_2$ is defined as:

$$f_2(D_1) = \frac{-\left(V_1 - C_1 \cdot \ln(D_1) - \dfrac{V_1}{D_1}\right)}{V_2 - C_2 \cdot \ln(f_1(D_1)) - \dfrac{V_2}{f_1(D_1)}} \qquad \text{in the } \textbf{GG} \text{ case}$$

and the resulting graph is given below.

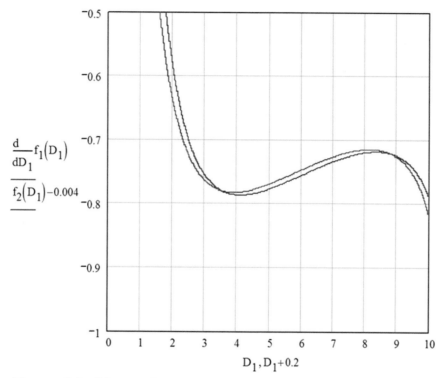

$$\frac{d}{dD_1}f_1(D_1)$$

$$\overline{f_2(D_1) - 0.004}$$

$$D_1, D_1 + 0.2$$

Figure 94. Slope of the GG Indifference Curve and Ratio $- P_1/P_2$

This confirms our contention that the ratio $- P_r/P_x$ is just another way of writing the first derivative of the market form of any of the true indifference curves derived in this work. Mathematicians would describe the solution process using the ratios $- P_r/P_x$ in lieu of the differential coefficients of the marginal rate of substitution of each pair of goods as a vector-space morphism. This is similar to the D operator method of solving differential equations.

These curves are based on the short-term demand function for competitive goods and the demand functions for speculative and **Giffen** goods derived earlier. The 6 graphs could therefore be regarded as just examples. However any equation for an indifference curve derived from any demand function and the universally accepted equation 48 will necessarily have this property.

When there is only one good involved in a market no substitution can possibly occur in that market. We have already evaluated the conditions for substitution between **2** goods when we considered partial equilibrium. We will now evaluate the conditions for **3,4,5,…,n** goods as follows.

Though there is no obvious a priori way to evaluate $\dfrac{dV}{dD_r}$ we can use

equation 91 to evaluate the vector $\underline{V}$ for **n = 3, 4, 5……** goods as follows:

when n = 3

$$
\begin{pmatrix}
1 & \dfrac{-P_1}{P_2} & \dfrac{-P_1}{P_3} \\[2ex]
\dfrac{-P_2}{P_1} & 1 & \dfrac{-P_2}{P_3} \\[2ex]
\dfrac{-P_3}{P_1} & \dfrac{-P_3}{P_2} & 1
\end{pmatrix}
\cdot
\begin{pmatrix} P_1 \\ P_2 \\ P_3 \end{pmatrix}
=
\begin{pmatrix} -P_1 \\ -P_2 \\ -P_3 \end{pmatrix}
$$

when n = 4

$$
\begin{pmatrix}
1 & \dfrac{-P_1}{P_2} & \dfrac{-P_1}{P_3} & \dfrac{-P_1}{P_4} \\[2ex]
\dfrac{-P_2}{P_1} & 1 & \dfrac{-P_2}{P_3} & \dfrac{-P_2}{P_4} \\[2ex]
\dfrac{-P_3}{P_1} & \dfrac{-P_3}{P_2} & 1 & \dfrac{-P_3}{P_4} \\[2ex]
\dfrac{-P_4}{P_1} & \dfrac{-P_4}{P_2} & \dfrac{-P_4}{P_3} & 1
\end{pmatrix}
\cdot
\begin{pmatrix} P_1 \\ P_2 \\ P_3 \\ P_4 \end{pmatrix}
=
\begin{pmatrix} -2 \cdot P_1 \\ -2 \cdot P_2 \\ -2 \cdot P_3 \\ -2 \cdot P_4 \end{pmatrix}
$$

when n = 5

$$
\begin{pmatrix}
1 & \dfrac{-P_1}{P_2} & \dfrac{-P_1}{P_3} & \dfrac{-P_1}{P_4} & \dfrac{-P_1}{P_5} \\[2ex]
\dfrac{-P_2}{P_1} & 1 & \dfrac{-P_2}{P_3} & \dfrac{-P_2}{P_4} & \dfrac{-P_2}{P_5} \\[2ex]
\dfrac{-P_3}{P_1} & \dfrac{-P_3}{P_2} & 1 & \dfrac{-P_3}{P_4} & \dfrac{-P_3}{P_5} \\[2ex]
\dfrac{-P_4}{P_1} & \dfrac{-P_4}{P_2} & \dfrac{-P_4}{P_3} & 1 & \dfrac{-P_4}{P_5} \\[2ex]
\dfrac{-P_5}{P_1} & \dfrac{-P_5}{P_2} & \dfrac{-P_5}{P_3} & \dfrac{-P_5}{P_4} & 1
\end{pmatrix}
\cdot
\begin{pmatrix} P_1 \\ P_2 \\ P_3 \\ P_4 \\ P_5 \end{pmatrix}
=
\begin{pmatrix} -3 \cdot P_1 \\ -3 \cdot P_2 \\ -3 \cdot P_3 \\ -3 \cdot P_4 \\ -3 \cdot P_5 \end{pmatrix}
$$

when n = 6

$$
\begin{pmatrix}
1 & \dfrac{-P_1}{P_2} & \dfrac{-P_1}{P_3} & \dfrac{-P_1}{P_4} & \dfrac{-P_1}{P_5} & \dfrac{-P_1}{P_6} \\[2mm]
\dfrac{-P_2}{P_1} & 1 & \dfrac{-P_2}{P_3} & \dfrac{-P_2}{P_4} & \dfrac{-P_2}{P_5} & \dfrac{-P_2}{P_6} \\[2mm]
\dfrac{-P_3}{P_1} & \dfrac{-P_3}{P_2} & 1 & \dfrac{-P_3}{P_4} & \dfrac{-P_3}{P_5} & \dfrac{-P_3}{P_6} \\[2mm]
\dfrac{-P_4}{P_1} & \dfrac{-P_4}{P_2} & \dfrac{-P_4}{P_3} & 1 & \dfrac{-P_4}{P_5} & \dfrac{-P_4}{P_6} \\[2mm]
\dfrac{-P_5}{P_1} & \dfrac{-P_5}{P_2} & \dfrac{-P_5}{P_3} & \dfrac{-P_5}{P_4} & 1 & \dfrac{-P_5}{P_6} \\[2mm]
\dfrac{-P_6}{P_1} & \dfrac{-P_6}{P_2} & \dfrac{-P_6}{P_3} & \dfrac{-P_6}{P_4} & \dfrac{-P_6}{P_5} & 1
\end{pmatrix}
\cdot
\begin{pmatrix} P_1 \\ P_2 \\ P_3 \\ P_4 \\ P_5 \\ P_6 \end{pmatrix}
=
\begin{pmatrix} -4 \cdot P_1 \\ -4 \cdot P_2 \\ -4 \cdot P_3 \\ -4 \cdot P_4 \\ -4 \cdot P_5 \\ -4 \cdot P_6 \end{pmatrix}
$$

etc. This sequence shows that the coefficients of every term in the vector **V**, for all values of **n,** is always given by – **(n – 2)**. The value of the terms in **V** for good **x** is then given by – **(n – 2).P_x** and for money as – **(n – 2)**. The vector **V** is therefore given by:

$$\underline{V} = -(n - 2).\underline{P} \qquad\qquad \text{Equation 93}$$

Because equations 88 are linearly independent and therefore non-singular, the matrix **D** is in principle invertible. Though MATHCAD cannot invert a general matrix of the form given above it can invert such a matrix for **n = 3, 4, 5, 6…**goods producing the results shown below:

when n = 3

$$
D^{-1} =
\begin{pmatrix}
0 & \dfrac{-1}{2} \cdot \dfrac{P_1}{P_2} & \dfrac{-1}{2} \cdot \dfrac{P_1}{P_3} \\[3mm]
\dfrac{-1}{2} \cdot \dfrac{P_2}{P_1} & 0 & \dfrac{-1}{2} \cdot \dfrac{P_2}{P_3} \\[3mm]
\dfrac{-1}{2} \cdot \dfrac{P_3}{P_1} & \dfrac{-1}{2} \cdot \dfrac{P_3}{P_2} & 0
\end{pmatrix}
$$

when n = 4

$$
D^{-1} =
\begin{pmatrix}
\dfrac{1}{4} & \dfrac{-1}{4} \cdot \dfrac{P_1}{P_2} & \dfrac{-1}{4} \cdot \dfrac{P_1}{P_3} & \dfrac{-1}{4} \cdot \dfrac{P_1}{P_4} \\[3mm]
\dfrac{-1}{4} \cdot \dfrac{P_2}{P_1} & \dfrac{1}{4} & \dfrac{-1}{4} \cdot \dfrac{P_2}{P_3} & \dfrac{-1}{4} \cdot \dfrac{P_2}{P_4} \\[3mm]
\dfrac{-1}{4} \cdot \dfrac{P_3}{P_1} & \dfrac{-1}{4} \cdot \dfrac{P_3}{P_2} & \dfrac{1}{4} & \dfrac{-1}{4} \cdot \dfrac{P_3}{P_4} \\[3mm]
\dfrac{-1}{4} \cdot \dfrac{P_4}{P_1} & \dfrac{-1}{4} \cdot \dfrac{P_4}{P_2} & \dfrac{-1}{4} \cdot \dfrac{P_4}{P_3} & \dfrac{1}{4}
\end{pmatrix}
$$

when n = 5

$$D^{-1} = \begin{pmatrix} \dfrac{1}{3} & \dfrac{-1}{6}\cdot\dfrac{P_1}{P_2} & \dfrac{-1}{6}\cdot\dfrac{P_1}{P_3} & \dfrac{-1}{6}\cdot\dfrac{P_1}{P_4} & \dfrac{-1}{6}\cdot\dfrac{P_1}{P_5} \\[2ex] \dfrac{-1}{6}\cdot\dfrac{P_2}{P_1} & \dfrac{1}{3} & \dfrac{-1}{6}\cdot\dfrac{P_2}{P_3} & \dfrac{-1}{6}\cdot\dfrac{P_2}{P_4} & \dfrac{-1}{6}\cdot\dfrac{P_2}{P_5} \\[2ex] \dfrac{-1}{6}\cdot\dfrac{P_3}{P_1} & \dfrac{-1}{6}\cdot\dfrac{P_3}{P_2} & \dfrac{1}{3} & \dfrac{-1}{6}\cdot\dfrac{P_3}{P_4} & \dfrac{-1}{6}\cdot\dfrac{P_3}{P_5} \\[2ex] \dfrac{-1}{6}\cdot\dfrac{P_4}{P_1} & \dfrac{-1}{6}\cdot\dfrac{P_4}{P_2} & \dfrac{-1}{6}\cdot\dfrac{P_4}{P_3} & \dfrac{1}{3} & \dfrac{-1}{6}\cdot\dfrac{P_4}{P_5} \\[2ex] \dfrac{-1}{6}\cdot\dfrac{P_5}{P_1} & \dfrac{-1}{6}\cdot\dfrac{P_5}{P_2} & \dfrac{-1}{6}\cdot\dfrac{P_5}{P_3} & \dfrac{-1}{6}\cdot\dfrac{P_5}{P_4} & \dfrac{1}{3} \end{pmatrix}$$

when n = 6

$$D^{-1} = \begin{pmatrix} \dfrac{3}{8} & \dfrac{-1}{8}\cdot\dfrac{P_1}{P_2} & \dfrac{-1}{8}\cdot\dfrac{P_1}{P_3} & \dfrac{-1}{8}\cdot\dfrac{P_1}{P_4} & \dfrac{-1}{8}\cdot\dfrac{P_1}{P_5} & \dfrac{-1}{8}\cdot\dfrac{P_1}{P_6} \\[2ex] \dfrac{-1}{8}\cdot\dfrac{P_2}{P_1} & \dfrac{3}{8} & \dfrac{-1}{8}\cdot\dfrac{P_2}{P_3} & \dfrac{-1}{8}\cdot\dfrac{P_2}{P_4} & \dfrac{-1}{8}\cdot\dfrac{P_2}{P_5} & \dfrac{-1}{8}\cdot\dfrac{P_2}{P_6} \\[2ex] \dfrac{-1}{8}\cdot\dfrac{P_3}{P_1} & \dfrac{-1}{8}\cdot\dfrac{P_3}{P_2} & \dfrac{3}{8} & \dfrac{-1}{8}\cdot\dfrac{P_3}{P_4} & \dfrac{-1}{8}\cdot\dfrac{P_3}{P_5} & \dfrac{-1}{8}\cdot\dfrac{P_3}{P_6} \\[2ex] \dfrac{-1}{8}\cdot\dfrac{P_4}{P_1} & \dfrac{-1}{8}\cdot\dfrac{P_4}{P_2} & \dfrac{-1}{8}\cdot\dfrac{P_4}{P_3} & \dfrac{3}{8} & \dfrac{-1}{8}\cdot\dfrac{P_4}{P_5} & \dfrac{-1}{8}\cdot\dfrac{P_4}{P_6} \\[2ex] \dfrac{-1}{8}\cdot\dfrac{P_5}{P_1} & \dfrac{-1}{8}\cdot\dfrac{P_5}{P_2} & \dfrac{-1}{8}\cdot\dfrac{P_5}{P_3} & \dfrac{-1}{8}\cdot\dfrac{P_5}{P_4} & \dfrac{3}{8} & \dfrac{-1}{8}\cdot\dfrac{P_5}{P_6} \\[2ex] \dfrac{-1}{8}\cdot\dfrac{P_6}{P_1} & \dfrac{-1}{8}\cdot\dfrac{P_6}{P_2} & \dfrac{-1}{8}\cdot\dfrac{P_6}{P_3} & \dfrac{-1}{8}\cdot\dfrac{P_6}{P_4} & \dfrac{-1}{8}\cdot\dfrac{P_6}{P_5} & \dfrac{3}{8} \end{pmatrix}$$

etc. This sequence shows that every term on the leading diagonal of the matrix **D^{-1}** for all values of **n** is always given by:

$$\frac{n-3}{2(n-2)}$$

in the

$$\lim_{n \to \infty} \frac{(n-3)}{2(n-2)} \to \frac{1}{2}$$

and if n is large

$$\frac{(n-3)}{2(n-2)} \approx \frac{1}{2}$$

Every other term in the matrix $\mathbf{D^{-1}}$ for all values of n is always given by:

$$\frac{1}{2(n-2)} \cdot \frac{-P_r}{P_X}$$

in the

$$\lim_{n \to \infty} \frac{-1}{2(n-2)} \cdot \frac{P_r}{P_X} \to 0$$

and if n is large

$$\frac{-1}{2(n-2)} \cdot \frac{P_r}{P_X} \approx 0$$

This system of simultaneous differential equations is therefore potentially a system of infinite dimensions. The inverted matrix $\mathbf{D^{-1}}$ can therefore be represented as:

$$D^{-1} = \frac{1}{2(n-2)} \cdot \begin{pmatrix} n-3 & \frac{-P_1}{P_2} & \frac{-P_1}{P_3} & \cdots & \frac{-P_1}{P_{n-1}} & \frac{-P_1}{P_n} \\ \frac{-P_2}{P_1} & n-3 & \frac{-P_2}{P_3} & \cdots & \frac{-P_2}{P_{n-1}} & \frac{-P_2}{P_n} \\ \frac{-P_3}{P_1} & \frac{-P_3}{P_2} & n-3 & \cdots & \frac{-P_3}{P_{n-1}} & \frac{-P_3}{P_n} \\ \cdot & \cdot & \cdot & \cdots & \cdot & \cdot \\ \cdot & \cdot & \cdot & \cdots & \cdot & \cdot \\ \cdot & \cdot & \cdot & \cdots & \cdot & \cdot \\ \cdot & \cdot & \cdot & \cdots & \cdot & \cdot \\ \frac{-P_n}{P_1} & \frac{-P_n}{P_2} & \frac{-P_n}{P_3} & \cdots & \frac{-P_n}{P_{n-1}} & n-3 \end{pmatrix}$$

Equation 94

or

$$D^{-1} = \frac{1}{2(n-2)} \cdot \begin{pmatrix} n-3 & \frac{d}{dD_1}D_2 & \frac{d}{dD_1}D_3 & \cdots & \frac{d}{dD_1}D_{n-1} & \frac{d}{dD_1}D_n \\[2mm] \frac{d}{dD_2}D_1 & n-3 & \frac{d}{dD_2}D_3 & \cdots & \frac{d}{dD_2}D_{n-1} & \frac{d}{dD_2}D_n \\[2mm] \frac{d}{dD_3}D_1 & \frac{d}{dD_3}D_2 & n-3 & \cdots & \frac{d}{dD_3}D_{n-1} & \frac{d}{dD_3}D_n \\[2mm] \vdots & \vdots & \vdots & & \vdots & \vdots \\[2mm] \frac{d}{dD_n}D_1 & \frac{d}{dD_n}D_2 & \frac{d}{dD_n}D_3 & \cdots & \frac{d}{dD_n}D_{n-1} & n-3 \end{pmatrix}$$

in the
$$\lim_{n \to \infty} \frac{1}{2(n-2)} \cdot \frac{d}{dD_r}D_x \to 0$$

and if n is large
$$\frac{1}{2(n-2)} \cdot \frac{d}{dD_r}D_x \approx 0 \qquad \text{Equations 95}$$

This innocent looking result is the coup de grace of the new theory of value and the nemesis of ordinal utility theory. The implication is that no substitution can take place between (in principle) any pair of goods in a general equilibrium since substitution can have no effect because all terms involving the marginal rate of substitution between all pairs of goods in the economy are reduced to zero. This concurs with the assumption concerning no substitution between goods that are not close substitutes in the new theory of value general equilibrium analysis and negates the assumption of ordinal utility theory that all goods are substituted for one another. It follows that the equilibrium price and level of demand of each good in the market is independent or autonomous of every other good in a general equilibrium, and that the 'natural' conditions for the equilibrium of supply and demand of each good in the market acting independently (as if no other goods existed) automatically satisfy the conditions for general equilibrium. Note that even though there can be no 'adjustment' in the market due to substitution there is still a process of 'demand side' adjustment because each consumer or buyer takes whatever quantity of a good he is able to buy with the resources he allocates to that good in accordance with his preferences and priorities.

The number of goods in a general equilibrium in even the poorest developing economy will, in principle, always be such that substitution is negligible. This would also apply to the consumption sets of almost all adults even in the poorest developing country and generally to the purchasing set of the smallest firm. We can demonstrate how the level of substitution falls as the number of goods in the economy or an individual consumer's or firm's purchasing set increases as shown in the graph below.

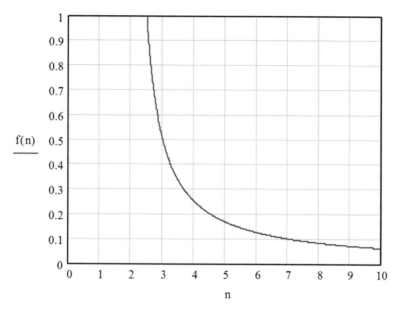

Figure 95. Proportion of Substitution Remaining as n Increases

Note that **f(n)** is actually a discrete relationship but it has been represented here as a continuous relationship to aid visualisation of the phenomenon.

With only 52 goods in the market as a whole or in any individual's or firm's purchasing set the level of substitution is reduced to **1%** of the level applying to each partial equilibrium between the same goods. It would seem that the only substantial group that would ever be affected by substitution in the general equilibrium of their own consumption sets are children purchasing goods from their own pocket money. Ordinal utility theory that was once thought to be universally applicable is now reduced to the economics of the sweetshop.

This may explain why their parents disposing of very much greater budgets can make decisions about all but their most expensive purchases very quickly while children seem to take an inordinate amount of time choosing between the cheapest sweets. The reason is that their parents face a much simpler problem since they do not need to 'optimise' most of their purchases because no substitution can take place between most of the goods in their consumption sets. Children on the other hand have a much more difficult problem partly because they are less experienced but also because they need to 'optimise' all their purchases since substitution will always be a significant factor to consider given their very small consumption sets. Furthermore, since their 'independent' income is very low and all their purchases are relatively cheap they have much less scope for approximation of optimum conditions by concentrating their attention on their more expensive purchases as their parents do. Why the condition shown above to affect all goods has now been relaxed so as to affect only most goods will be explained later when we consider virtual homogeneous markets.

From equations 91, 92, 93 and 94 the general equilibrium of the entire market is then given by:

$$
\underline{P} = \frac{1}{2(n-2)} \cdot
\begin{pmatrix}
n-3 & \frac{-P_1}{P_2} & \frac{-P_1}{P_3} & \cdots & \frac{-P_1}{P_{n-1}} & \frac{-P_1}{P_n} \\
\frac{-P_2}{P_1} & n-3 & \frac{-P_2}{P_3} & \cdots & \frac{-P_2}{P_{n-1}} & \frac{-P_2}{P_n} \\
\frac{-P_3}{P_1} & \frac{-P_3}{P_2} & n-3 & \cdots & \frac{-P_3}{P_{n-1}} & \frac{-P_3}{P_n} \\
\cdot & \cdot & \cdot & \cdots & \cdot & \\
\cdot & \cdot & \cdot & \cdots & \cdot & \\
\cdot & \cdot & \cdot & \cdots & \cdot & \\
\cdot & \cdot & \cdot & \cdots & \cdot & \\
\frac{-P_n}{P_1} & \frac{-P_n}{P_2} & \frac{-P_n}{P_3} & \cdots & \frac{-P_n}{P_{n-1}} & n-3
\end{pmatrix}
\cdot -(n-2) \cdot
\begin{bmatrix}
P_1 \\ P_2 \\ P_3 \\ \cdot \\ \cdot \\ \cdot \\ \cdot \\ P_n
\end{bmatrix}
$$

Equation 96

Even though each term in the matrix **D^{-1}** is 'amplified' by **– P$_x$/2** the condition of no substituting in a general equilibrium still holds for the reasons discussed when reviewing the linear independence of equations 88 and because any finite number multiplied by zero is still zero.

The equilibrium solution then simplifies to:

$$
\underline{P} =
\begin{bmatrix}
\frac{-(n-3)}{2} \cdot P_1 & \frac{P_1}{2} & \frac{P_1}{2} & \cdots & \frac{P_1}{2} & \frac{P_1}{2} \\
\frac{P_2}{2} & \frac{-(n-3)}{2} \cdot P_2 & \frac{P_2}{2} & \cdots & \frac{P_2}{2} & \frac{P_2}{2} \\
\frac{P_3}{2} & \frac{P_3}{2} & \frac{-(n-3)}{2} P_3 & \cdots & \frac{P_3}{2} & \frac{P_3}{2} \\
\cdot & \cdot & \cdot & \cdots & \cdot & \\
\cdot & \cdot & \cdot & \cdots & \cdot & \\
\cdot & \cdot & \cdot & \cdots & \cdot & \\
\cdot & \cdot & \cdot & \cdots & \cdot & \\
\frac{P_n}{2} & \frac{P_n}{2} & \frac{P_n}{2} & \cdots & \frac{P_n}{2} & \frac{-(n-3)}{2} \cdot P_n
\end{bmatrix}
$$

Because there are **n − 1** terms that are not on the leading diagonal of each row of this matrix, the row sum for every row in the matrix and all values of n is always given by:

$$\frac{-(n-3)}{2} \cdot P_r + \frac{(n-1)}{2} \cdot P_r \quad \text{which simplifies to } P_r$$

This confirms that the condition for a partial equilibrium of every pair of goods in the economy holds for a general equilibrium since the expected solution for all **n** goods in the market is given by the set of unique prices in the vector **P**.

$$\underline{P} = \begin{pmatrix} P_1 \\ P_2 \\ P_3 \\ . \\ . \\ . \\ . \\ . \\ P_n \end{pmatrix}$$

Mathematicians should note that the method of solution used here has in principle a much wider applicability to the solution of any similarly symmetrical very large system of algebraic, differential or partial differential equations in which the dimensions of the matrix would have a similar effect on the coefficients of the inverse matrix and on the vector of dependent variables. Furthermore it is conjectured that only such systems of equations are soluble in very large systems.

Because the row sum is independent of the number of goods in the market the equilibrium solution will not only work for the entire market but for any arbitrarily defined sub-set of goods greater than 2. However, readers should remember that substitution only tends to zero when **n** is large, though for practical purposes substitution is reduced to **<1%** of the level arising in partial equilibrium when **n > 52**. As we have already shown that the partial equilibrium of an individual for any pair of goods is related to the partial equilibrium of those goods in the market as a whole, that partial equilibrium solutions are always unique and still hold in a general equilibrium and that each market for each good acts independently, the general equilibrium solution is therefore likely to apply equally to all individual's consumption set and to the purchasing set of any individual firm.

The conclusions reached above are materially different from those widely held to apply in a general equilibrium. The presumption has been that market adjustment, **Walras'** process of 'tatonnement', produces a result tending to equilibrium. It is now evident why **Walras** could not suggest any practically believable process of adjustment. There is no such process. The whole market in aggregate automatically gets it 'right first time' when each market, for each good, acting independently, achieves its own equilibrium of supply and demand.

For a short-term general equilibrium there can be only two independent variables in the market for any good. Each buyer decides how much money he wishes to spend on each good and each seller decides on the amount they each wish to sell of each good. Note the use of the word amount applies rather than quantity because we have specifically excluded discrete demand from our analysis for the time being.

Effective demand is therefore constrained to be not greater than the available supply. For a market to clear, demand is also constrained to be not less than the available supply. In any general equilibrium, demand is thus constrained to be equal to the available supply.

However, for a speculative or **Giffen** good market to be sustained, the demand for each good must be consistent with the demand function applicable to such a market in each case. If for example we assume the demand functions of equations 35 and 43 apply then for a speculative good the total value of consumption or use is given by:

$$V_S = C_D \cdot D_S \cdot \ln(D_S) + V_C$$

Where
V_S is the total value of consumption or use of a good in a speculative market

V_C is the total value of consumption or use of the same good in a competitive market

C_D is the direct cost of the good

D_S is the demand for the good in a speculative market

Then if the available supply is greater than:

$$D_S = \frac{1}{C_D} \cdot \frac{(V_S - V_C)}{W\left[\frac{1}{C_D} \cdot (V_S - V_C)\right]}$$

Equation 97

the market for a speculative good will begin to revert to the conditions of a competitive market.

For a **Giffen** good the total value of consumption or use is given by:

$$V_G = V_C \cdot D_G - C_D \cdot D_G \cdot \ln(D_G) - V_C$$

Where

V_G is the total value of consumption or use of a good in a **Giffen** good market

V_C is the total value of consumption or use of the same good in a competitive market

C_D is the direct cost of the good

D_G is demand for the good in a **Giffen** good market

Then, if the available supply is greater than:

$$D_G = \exp\left[\left[W\left[\frac{-(V_G + V_C)}{C_D} \cdot \exp\left(\frac{-V_C}{C_D}\right)\right] + \frac{V_C}{C_D}\right]\right] \qquad \text{Equation 98}$$

the market for a **Giffen** good will begin to revert to the conditions of a competitive market. In all possible cases therefore attainment of a general equilibrium implies that effective demand for every good is subject to an equality constraint, or, put more simply, effective demand for every good is fixed. If demand is fixed then so too is price since price is determined by the demand function for each good that is applicable to the prevailing conditions in that market. Note this even applies to monopoly and oligopoly supplied goods for the markets to clear. Hence demand for each good is independent of one another and thus the price of each good is also.

The solution derived above expresses the general equilibrium problem in terms of price because historically that is how general equilibrium was originally expressed. We could however have just as easily expressed the result in terms of demand as is the customary way to express a partial equilibrium. All that is required to achieve this is to differentiate equation 84 by P_r instead of D_r. All of the resulting analysis and equations would be identical except that demand would replace price, the marginal rate of price adjustment or relative change of price would replace the marginal rate of substitution, a vector of demand **D** would replace the vector of prices **P**, the vector **V** would contain differential coefficients derived with respect to the rate of change of price of each good **r** and the matrix **D** and its inverse would be replaced by a matrix **P** and its inverse of differential coefficients giving the marginal rate of price adjustment or relative change of price with respect to a change in the price of each good **r**. Each term in this matrix (other than terms on the leading diagonal) would then be equal to $- D_r/D_x$. This system defines each firm's or seller's behaviour as opposed to each consumer's or buyer's.

Just as all terms (other than terms on the leading diagonal) of the inverse matrix D^{-1} tend to zero as the number of goods in the economy **n** tends to ∞ so all terms (other than terms on the leading diagonal) of the inverse matrix

P^{-1} tend to zero as **n** tends to ∞ Therefore not only can there be no substitution in a general equilibrium but also there can be no price adjustment due to marginal relative changes of price either.

This however does not imply no adjustment of prices or levels of demand takes place, just that the mechanism of adjustment is not as is supposed in ordinal utility theory. Firms, or other sellers, must adjust their prices for the market to clear in the same way that consumers, or other buyers, must adjust their levels of demand. In both cases the adjustment must ensure $P_x.D_x = V_x$ irrespective of the form of demand function or market structure. Adjustment is achieved because the only variables in the demand side model are prices that are adjusted in the supply side of the system, and the only variables in the supply side model are levels of demand that are adjusted in the demand side of the system.

The adjustment of demand is automatic since consumers will take whatever quantity of each good **x** their allocation to **x** allows them to buy at prevailing market prices. The adjustment of price however requires a conscious decision by all firms that fail to sell their output (or sell it too quickly) using the rules for price adjustment described below.

The General Equilibrium of Real Markets

Now we have established the mathematical conditions for general equilibrium we can consider the practical conditions. We do not need to insist that no trading takes place under dis-equilibrium conditions, as **Walras** and others have suggested, so long as equilibrium conditions are established reasonably expeditiously. This is likely because as we have seen the market for each good acts independently. Neither must we insist on absolute optimum conditions, provided consumers and firms can achieve a reasonable approximation to optimum conditions by the methods described earlier in this work.

Markets will clear more or less provided firms offer their products at reasonably appropriate prices because consumers and, now by analogy, all buyers will normally allocate their resources to competing uses and take whatever amount they are able to buy for that amount of money. Firms also do this formally through their budgets particularly for their indirect purchases. For directly derived material and labour demands, firms must flex their level of purchases more or less in proportion to their output. Though this form of behaviour will be more common for a firm than a consumer, even consumers will behave in a similar way with respect to complementary goods.

Though the system of equations includes all goods, in monopoly markets the prices and volumes offered for sale are not determined by the market but by the monopolists. Monopoly markets will therefore only clear if monopolists choose compatible combinations of price and volume. The appropriate set of 'equilibrium' prices and volumes for goods in such a market, one pair of which each monopolist must adopt if they seek to clear all that they produce, is bounded by the quadratic solutions for both monopoly price and volume and the competitive solution.

Similarly prices and volumes in oligopoly markets are not determined by the market but by the interaction of the strategic behaviour of oligopoly firms. The appropriate 'equilibrium' price and volume for goods in such markets that each oligopolist must adopt if they seek to clear all that they produce is given by the general oligopoly price and volume solution.

Though monopoly and oligopoly supplied goods do not 'participate' in the 'natural' market equilibrium, the presence of any number of such goods in the market does not prevent a determinant solution to the system of equations 88 for all other goods because equilibrium applies equally to all arbitrarily defined sub-sets of goods greater than two. Strictly provided there are more than two goods traded competitively (in the conventional sense), speculatively or as **Giffen** goods their markets will still clear despite any number of goods supplied by monopolists or oligopolists.

Alternatively each overall matrix of **n** equations could, in principle, be reduced to **n − m − θ** equations with **n − m − θ** unknowns and **n − m − θ + 1** terms, where **m** is the number of goods traded under monopoly conditions and **θ** is the number of goods traded under oligopoly conditions. Identical solutions will result from either method because as we have seen the general equilibrium solution applies both to the market as a whole or any arbitrarily defined subset of 3 or more goods within the market. This is analogous to the **Walrasian** solution although it is not restricted to competitive goods.

Irrespective of the market conditions, all firms necessarily have to make price and volume decisions because generally no external mechanisms exist to tell them how much to produce and what prices to charge before they offer their goods to the market. Whatever is the final outcome and however powerless a firm may ultimately prove to be, no firm is a price taker initially. The price adjustment mechanism required of any firm that seeks to sell all that they produce is therefore essentially the same irrespective of whether they trade competitively (in the conventional sense), monopolistically or oligopolistically.

Provided each firm responds appropriately in adjusting prices if they fail to sell their output, this is all that is required to ensure that markets will clear more or less. We do not have to assume the **Walrasian** rule for price changes because firms will know if they are trading in a market with an inverse demand curve or if they don't they will soon find out. If firms are trading in such a market, they simply adopt the opposite rule.

In a sense, markets must always clear because of the circular flow of income. If a firm fails to sell their output then they themselves have effectively bought it. This of course is not the sense in which general equilibrium theory suggests that markets clear. It is however helpful to practical equilibrium in real markets because it implies absolute clearance is unnecessary. Most firms plan to hold inventory, either of finished products, materials, work in progress or other supplies. Similarly, service providers maintain 'reserves of capability'. Since sales fluctuate, firms can be expected to have a tolerance for increased stock or unused capacity before they reduce their prices. The presumption is that once this tolerance is breached, appropriate price changing decisions will be made. This may also apply if they sell their output more quickly than expected as well as more slowly. Alternatively a firm may simply choose instead to increase or decrease their levels of production and resources rather than to increase or decrease their prices.

As we have seen consumers can approximate an optimum allocation of resource by relatively simple means and it is presumed firms can do likewise using similar simple satisficing rules, particularly for trivial purchases. Firms only need to use more formal rational decision rules where an item of expenditure represents a significant proportion of their budget, for example for capital purchases. The budgetary control techniques used by firms work very much in this way. Note this behaviour is an exact parallel with consumers' behaviour in approximating optimum choice.

The only set of goods that we have so far not included in the new theory of value general equilibrium system is the very large set of goods with discrete demand. We know from the techniques of integer programming that an optimum integer solution is not necessarily the closest solution to the underlying continuous solution. If it were ever important to determine an optimum integer solution exactly among a set of close substitutes it would obviously be possible to define the optimisation process as an integer programming problem. These conditions might apply for example to large volume buyers in the motor industry or for a supermarket chain. In general however, this is unnecessary because most consumers and buyers usually only approximate optimisation for all but their most expensive purchases. Taking the integer solution closest to the underlying continuous solution of the present analysis is therefore generally quite sufficient. The techniques of integer programming show that such a solution will be very close to the absolute optimum in almost all cases. Furthermore absolute clearance of markets is not required because sales are buffered by stock. The set of goods and services to which the present analysis has therefore now been shown to apply includes all goods and services in all real markets (in the economic sense).

The general equilibrium solution derived from the new theory of value is much more mathematically specific (relying on quantifiable notions of preference and precise equations) than the **Walrasian** system that depends on general functional relations and the un-quantifiable variables **U** and **T**. Furthermore we have not just formed the equilibrium equations but shown how to solve them. The new system is also a more general solution because it does not depend on all markets being perfectly competitive and it specifically includes money and borrowing. The new theory is also more convincing since it is equally applicable to all real markets (in the economic sense) irrespective of their structure.

Furthermore, earlier neoclassical views that competitive markets are preferable and the norm in market economies now seems a much more problematic conclusion. Monopoly pricing has been shown to be much more widespread than was hitherto supposed due to the existence of virtual monopoly prices in speculative markets. The applicability of speculative market models to a wide cross section of markets, including, for example, the fashion market, also makes orthodox views less credible. The general pricing solution also implies that markets for branded goods will all involve monopoly or oligopoly pricing above competitive equilibrium levels.

The set of goods that are affected by monopoly pricing includes the natural monopolies of the public utilities. It is difficult to imagine a modern industrial society without power or piped water supplies, sewage systems or modern transportation or telecommunication systems. The Walrasian model is a gross over simplification that bears little comparison with real economies. General equilibrium under the new theory of value, on the other hand, is difficult to separate from the observable behaviour of the real economy.

We can demonstrate that the process defined above does actually work from the data used for the general equilibrium simulation shown in Appendix VI. This simulation was constructed using a Microsoft Excel spreadsheet giving the negative ratios of prices and levels of demand for **20** goods. All demand functions assumed are the simplest form of those derived in this work. Each of the first **19** goods is automatically randomly selected to be a competitive, speculative or **Giffen** good, to have continuous or discrete demand and to have a 'natural' equilibrium of price and demand or to have a randomly selected arbitrary price within the set of allowable prices under monopoly or oligopoly conditions. Monopoly and oligopoly suppliers are however assumed to have chosen compatible combinations of price and volume required for equilibrium. The last of the **20** goods is assumed to be money with a price of **1**. Money is placed last simply for convenience. The system would work just as well irrespective of whether money was included or not, and where it was placed. Note the simulation produces different data each time it is run. Each matrix so formed was then inverted in MATHCAD as two distinct sub-sets of **10** goods because a **10x10** matrix is the largest matrix that MATHCAD can display at one time in its entirety. The system was thence 'solved' in MATHCAD for price and could also be solved for demand. This confirms both that the resulting solution is an equilibrium system in terms of both price and demand when each market acts independently, and that the two arbitrary sub-sets of **10** goods can be correctly solved separately. This also confirms that the system works perfectly well whether or not one of the goods in the matrix is money. As we might expect, very slight errors in price will arise in some cases, particularly for discrete demand, but generally the equilibrium solution is very precise.

The simulation also shows that any change in the price and level of demand of a single good, though it affects an entire row and an entire column of the matrix **D** (and the matrix P when the problem is defined in terms of demand rather than price), affects only a single price (and level of demand) of the good concerned. This confirms that the markets for each good are independent of one another in a general equilibrium.

Of course, due to a change of preference (for example arising from a change of fashion) aggregate purchases of several goods can change together in a complementary or reciprocal manner. The simulation can also support this scenario. It is however important to realise that it is an interaction between buyer preferences (for complimentary goods for example) that causes such behaviour in the market, not an interaction within the market mechanism itself.

Virtual Homogeneous Markets and the Limits of Ordinal Utility Theory

The derivation of the general equilibrium solution clearly shows that there can be no substitution between 'any' pair of goods in a general equilibrium. That the process of substitution does take place between close substitutes is however equally certain. So how can these seemingly irreconcilable conditions be compatible?

This question may be answered by considering what constitutes a homogeneous market. Primary goods or goods such as timber, grain or crude oil of a particular type and quality can each reasonably be considered to all intents and purposes to be homogeneous, even though each batch or lot may have slight differences. Branded goods on the other hand, such as the set of small family cars, although designed to perform essentially similar tasks, differ in appearance, features and performance. Indeed, manufacturers of such goods deliberately seek to differentiate their products by such features as well as by branding and promotion. The only conceivable way we can square the circle and make substitution allowable between branded goods (for example) that are 'close' substitutes, whilst conforming with the result that no substitution can take place between goods in a general equilibrium, is to consider the markets for goods that are 'close' substitutes as homogeneous, despite any obvious differences between such goods.

This leads to the concept of a 'virtual homogeneous market'. Goods within a virtual homogeneous market, though distinct, behave as if they were all a single homogeneous good with respect to all other goods, while each behaves as a differentiated good with respect to other goods within the virtual homogeneous market. Therefore while no substitution will take place between virtually homogeneous goods and other goods, substitution will take place between goods within a virtual homogeneous market. The notion of a virtually homogeneous market is thus very similar to the marketing concept of a market sector, but not quite the same as the economic concept of a market segment.

Because the number of differentiated goods within a virtual homogeneous market must by definition be small (otherwise substitution would once again be negligible), substitution will be a significant factor to take into account in the approximation of optimum choice pursued by each consumer or firm within that market. The principles of ordinal utility will therefore apply and choice will approximate the rules of that theory. Note however that the perceived size of each buyer's set of close substitutes is limited by the extent of their search, discussed earlier, and by any characteristic of the good in question they each define as essential. Substitution will therefore occur very much more often than the actual number of goods in a virtual homogeneous market implies. For a general equilibrium between a set of virtually homogeneous goods and all other goods or sets of virtually homogeneous goods, the principles of ordinal utility cannot apply and choice must be based on the principles of the new theory of value. Ordinal utility theory therefore though not entirely abandoned or overturned now has a much more limited applicability.

Factor Markets

The markets for the factors of production differ from other markets because the supply to these markets is not in the short term produced by the economic process but exogenously. In aggregate the supply of land is determined by geological processes, the supply of labour by reproductive processes and the supply of capital by historical processes. The amount of each available at any point in time is not that which the market demands but just that which is available to be used.

These markets only clear in a very restricted sense. They clear because neoclassical general equilibrium analysis defines the supply of the factors of production not as all that is available, but as only that which is sold at prevailing factor prices. Any that are not sold are assumed not to have been supplied. In fact, this is merely a 'get out'. The supply is always that, which is available irrespective of whether it is sold and used or not, otherwise such markets could not be said to clear. One does not have to look very hard to find examples of idle resources in any firm, or of all the factors of production in the economy as a whole. Neoclassical General equilibrium analysis thus turns its back on the waste of the factors of production that are unused. Pareto efficiency is therefore an illusion because it ignores the inefficient waste of unused factors of production.

The orthodox view is to assume that only those factors whose marginal returns exceed their cost should be used and any that remain that are less productive are simply not taken up. This formulation only works if a sub-optimal definition of cost is used. The marginal opportunity cost of unused capital goods owned by a firm or un-worked land available to be farmed in hand (excluding set aside) or of either on a long-term lease is always zero because, at least in the short term, they are sunk costs. The marginal productivity of such assets if they can produce anything at all would therefore be infinite.

So how does this paradox arise? What causes a firm not to use all the factors of production that are available to it is an incorrect perception of their true cost. Again the orthodox view of economists is to notionally charge for the use of assets at their economic rent in a way first suggested by (**David Ricardo**, 1817). Accountants also have a similarly incorrect, though differently formulated, perception of costs. Both formulations, result in an over estimation of the cost of a marginal asset. Applying **Ricardo's** formulation, when such an asset is used it is assumed it should carry an economic rent, whereas when it is not used it does not. In fact, its cost remains unchanged whether it is used or not.

The same is true of unemployed labour that also has an economic rent. Furthermore idle labour still has to be funded by social security benefits paid out of taxation. The cost of all benefits falls ultimately on firms (including sole traders and subsistence producers) because they are the only producers. The lowest rates of pay existing in the labour market differ little from that which is available from the benefit system. Were this not so there would be no such thing as a poverty trap.

A better way of treating these issues is to use another formulation of **David Ricardo's**, the concept of comparative advantage. Using this approach, usually reserved just for international trade, all resources would have comparative advantage in producing something, even though many would not have an absolute advantage in producing anything. Such a treatment would allow far fewer (and possibly no) assets and resources to remain unused. It would also help us to better understand how the economy works.

The orthodox formulation of general equilibrium analysis is to assume the result is a long-term equilibrium. All firms are therefore assumed to be technically efficient because it is held that, in the long-term, only the technically efficient can survive. The long-term in this sense could be very long indeed satisfying the **Keynesian** definition of the long term in which we (and our grandchildren) are all dead.

The formulation of the new theory of value is to assume a short-term equilibrium that does not depend on technical efficiency. Indeed using the principle of comparative advantage allows the possibility of the protracted coexistence of firms of markedly different efficiency, and the use of all the factors of production of markedly different productivity. The long-term coexistence of different levels of performance is a far more realistic model of observable reality than the assumption of technical efficiency of orthodox neoclassical theory.

Chapter 12. Welfare Implications of the New Theory of Value

Just as we can define an individual's total utility function as:

$$u = \sum_{x=1}^{n} P_x q_x$$

so, we can define the aggregate utility function as:

$$U = \sum_{x=1}^{n} P_x D_x$$

This may give the reader some difficulty because it entails the addition of what have hitherto been thought of as personally defined utilities across all individuals in the economy, implying inter-personal comparisons of utility. We need therefore to consider carefully why this is so because all that follows depends on this result.

We defined an individual's total utility from his consumption function because all the consumption functions for competitive goods are saturating functions demonstrating satiation of human wants and therefore diminishing marginal utility.

Thus
$$v = \sum_{x=1}^{n} P_x q_x$$

can also be written as
$$u = \sum_{x=1}^{n} P_x q_x$$

We have shown that the aggregate consumption of a good is formed by the vector addition in two-dimensional vector space of the consumption functions of all individuals in the economy for that good. This is because each consumer will receive a different income and therefore, even apart from any differences of preference, have a different effect on overall consumption of each good.

Thus:

$$\begin{pmatrix} V_x \\ Y \end{pmatrix} = \begin{bmatrix} \sum_{i=1}^{N} \left(\alpha_x \cdot y \cdot \exp\left(\dfrac{-y}{\lambda}\right) + \beta_x \right) \\ \sum_{i=1}^{N} y \end{bmatrix}$$

We defined the aggregate preference a_x for good x for all consumers from:

$$a_x \cdot Y \cdot \exp\left(\frac{-Y}{N\lambda}\right) = \sum_{i=1}^{N} \alpha_x \cdot y \cdot \exp\left(\frac{-y}{\lambda}\right)$$

and therefore
$$a_x = \frac{\displaystyle\sum_{i=1}^{N} \alpha_x \cdot y \cdot \exp\left(\frac{-y}{\lambda}\right)}{\left(Y \cdot \exp\left(\frac{-Y}{N\lambda}\right)\right)}$$

The parameter a_x is thus the income weighted average preference of all consumers for good x. '**a_x**' therefore does not distort the different 'contribution' each consumer makes to overall consumption due to his differing income and preferences.

We defined the aggregate priority of good **x** for all consumers as:

$$b_x = \sum_{i=1}^{N} \beta_x$$

just the sum of individual 'priorities' because β_x is independent of income.

Thence when we defined the short-term competitive demand function for good x in equations 18 and 19 as:

$$D_x = (a_x Y e^{-Y/\lambda} + b_x)/ P_x$$

or
$$P_x = (a_x Y e^{-Y/\lambda} + b_x)/ D_x$$

we effectively preserved each individual's consumption function and therefore his utility function, preferences and priorities in the term **$a_x Y e^{-Y/\lambda} + b_x$**.

When we were considering the value system of all consumers we noted that each consumer used the opportunity cost of obtaining or retaining a good as his basis of valuing that good. Each consumer therefore simply adopts the market price of a good as his measure of its value. They have no choice about this because all consumers are almost invariably price takers. It is also essential that they adopt this system of value so that they correctly economise in the use of all the goods in their consumption set. This is why the individual's consumption and demand functions for a good includes the overall market price **P_x** but only his individual quantity demanded and consumed **q_x**. Note also that **q_x** is a function of a consumer's own personal preference **α_x** and priority **β_x** for good **x** and that these parameters are behavioural parameters because they are learned behaviour.

When we considered partial equilibrium we showed that the price ratio – P_1/P_2 naturally followed the slope of the market indifference curve, a form of aggregate indifference curve for all consumers, that all consumers have identical indifference functions, having identical indifference functions does not imply identical indifference maps. We also showed that utility or value were distinct from preference and that the equilibrium solution was necessarily unique for all possible combinations of competitive, speculative and **Giffen** goods. Equilibrium under the new theory of value and under ordinal utility theory have an equivalent representation under the two systems, and that under the new theory of value all consumers value economic goods as the market values them.

When we considered general equilibrium we showed that the unique partial equilibrium solution for each pair of goods also held in a general equilibrium and that substitution could not occur in a general equilibrium. This is inconsistent with the assumptions of ordinal utility theory but one of the assumptions of the new theory of value. All consumers must therefore value all goods in a general equilibrium at their market value. Thus utility and value are just two different ways of expressing the same thing.

These equations confirmed that no substitution can take place between goods in a general equilibrium because all terms involving substitution in the substitution matrix **D** are reduced to zero in the inverted substitution matrix $\mathbf{D^{-1}}$ This confirms the principal assumption of the new theory of value, initially set aside when considering general equilibrium, that substitution between goods in a general equilibrium, that are not close substitutes, is negligible tending to zero.

Since each individual must subscribe to the same system of values to correctly economise in the use of all the goods he consumes, the total consumption function and therefore the total utility function of every consumer is valued identically. Since all utility functions are valued identically there is no difficulty in adding them together even where this entails adding different individual's utility functions.

An individual's total consumption function written in the form:

$$v = \alpha y e^{-y/\lambda} + \beta$$

also defines his total utility function:

$$u = \alpha y e^{-y/\lambda} + \beta$$

and an individual's aggregate marginal propensity to consume:

$$\frac{dv}{dy} = \alpha e^{-y/\lambda}(1 - y/\lambda)$$

also defines his marginal utility of income:

$$\frac{du}{dy} = \alpha e^{-y/\lambda} (1 - y/\lambda)$$

Equations of identical form were also derived for aggregate consumption, utility and marginal utility of income.

We have also shown that though β that defines an individual's level of borrowing (or lending), can take any value, positive or negative, net aggregate borrowing **B** in a general equilibrium, though subject to noise in macro-economic data, must effectively be zero. Since the value of λ is related to the value of money and thus cannot be a behavioural parameter λ must therefore be constant for all individuals. It follows that if we know an individual's income, consumption (income minus savings including pensions and life assurance) and borrowing (or lending) we can calculate his value of α and β. We can then calculate his total utility and marginal utility of income very precisely.

Note that α and β represent an individual's aggregate preference and priority for all his consumption in his own estimation so that the values of total and marginal utility that result are Pareto optimal. All this information is available to the government or the banks and other financial institutions. With suitable legislation, all the necessary information to maximise social welfare could be collected by government.

This then provides a means by which society can maximise its utility. The condition for maximising utility across the whole economy is achieved when the marginal utility of income of all consumers is equal, implying an equal distribution of income. This condition will yield significantly higher total utility for society than the Pareto efficient allocation (which only prevents loss of value) since, ceteris paribus, the higher a person's income the more his total consumption will satiate and the lower his marginal utility of income will be.

Because the distribution of income and initial endowments of wealth are skewed, great benefit to society will result if income and wealth were redistributed to achieve this condition. This would both increase society's total utility significantly and be compatible with the greatest happiness principle. Note that the extent to which a society fulfils the greatest happiness principle can be measured for all competitive goods using the result of equation 30 as:

$$V_{ST} = \sum_{x=1}^{n} V_{SX}$$

$$V_{ST} = \sum_{x=1}^{n} (V_x \log_e D_x - V_x) \qquad \text{Equation 99}$$

where V_{ST} is the total 'value' of 'consumer surplus' derived from all competitive goods

x is an arbitrary competitive good

and n is the total number of competitive goods in the economy

You may recall the definition of a competitive good used in this essay implies

$$\frac{dP_x}{dD_x} < 0$$

Goods that do not satisfy this condition cannot have a 'consumer surplus' so equation 99 gives the total 'value' of consumer surplus for all goods other than the first of each such good in the economy, for which consumer surplus is always indeterminate.

We can calculate the effect of an equal distribution of income on total utility as follows. The matrix below shows the observed actual distribution of household income, in bands of **£10,000**, derived from the (Office for National Statistics, 2004) 'Expenditure and Food Survey 1999 – 2000' shown in Appendix IV Table 14. It would be better to base our analysis on earned income but since rather surprisingly the proportion of unearned income does not increase as income rises little distortion will result if we use gross income.

MATHCAD provides a function for plotting graphs from tabulated data represented in the form of a matrix. The figures in the first column of the matrix below are taken from the first row labelled 'Gross households' in the main body of table 14 and the 'mean or middle income' of each band of income shown in the header band of the table are entered in the second column of the matrix,

$$\text{Coords} = \begin{pmatrix} 7023000 & 5000 \\ 6079000 & 15000 \\ 4793000 & 25000 \\ 3173000 & 35000 \\ 1866000 & 45000 \\ 1024000 & 55000 \\ 511000 & 65000 \\ 277000 & 75000 \\ 201000 & 85000 \\ 85000 & 95000 \\ 304000 & 105000 \end{pmatrix}$$

Two variables are then defined in MATHCAD as follows:

$$n = \text{Coords}^{\langle 0 \rangle}$$

and $$y = \text{Coords}^{\langle 1 \rangle}$$

Where   n is the estimated gross number of households in each income band,

and   y is the midpoint of income in each band

We can then plot the actual distribution of income as shown below. We have used this method of plotting the distribution rather than the more conventional histogram or frequency polygon because our analysis requires a detailed comparison of actual and theoretical distributions. This will be much simpler when the actual distribution is represented as a 'continuous' curve. Fortunately, MATHCAD provides the tools to enable us to do this.

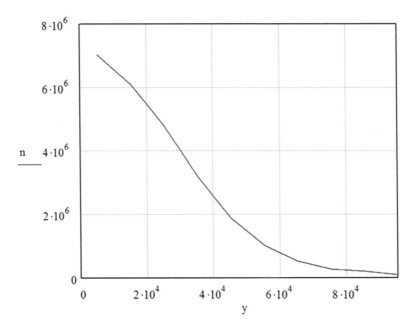

Figure 96. Distribution of Household Income in The UK 1999 – 2000

If we define an index variable **i** to represent rows of the matrix, counting from the bottom, we can then calculate the mean μ and standard deviation σ for the actual distribution of household income as shown below. Note that the MATHCAD default is to assume the first row and column of the matrix has an index of **0** not **1** as might be expected.

$$\mu = \frac{\displaystyle\sum_{i=0}^{b-1} n_i \cdot y_i}{\displaystyle\sum_{i=0}^{b-1} n_i}$$

and

$$\sigma = \sqrt{\dfrac{\displaystyle\sum_{i=0}^{b-1} n_i \cdot \left(y_i - \mu\right)^2}{\displaystyle\sum_{i=0}^{b-1} n_i}}$$

Where   b is the number of bands of income in the Table

The resulting value for the mean μ and standard deviation σ obtained from the data in the matrix on page 211 derived from the (Office for National Statistics, 2004) is **£24,019** and **£19,841** respectively, when calculated by the weighted mean and standard deviation method.

Clearly the distribution in figure 96 is severely skewed. It is often said that incomes are distributed log normally. If we plot a log normal distribution with these values of the parameters μ and σ it shows that although the actual distribution of household income is indeed severely skewed it is not quite as skewed as would be the case if incomes were distributed log normally, particularly at higher values of f(y).

This is evident from the graph shown below plotted for a log normal distribution with the same parameter values from the equation:

$$n = \dfrac{N}{y \cdot \ln(\sigma) \cdot \sqrt{2\pi}} \cdot \exp\left[\dfrac{-1}{2 \cdot (\ln(\sigma))^2} \cdot (\ln(y) - \ln(\mu))^2 \right]$$

where   N is the total number of households in the economy

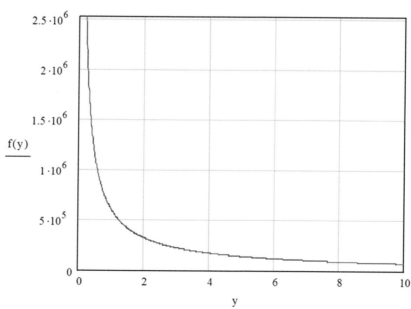

Figure 97. Income Distribution Assuming a Log Normal Distribution

Using the parameters defined above we can calculate aggregate household income **Y** from the equation:

$$Y = \mu \cdot N \qquad \text{Equation 100}$$

Where   N is the total number of households

From the value of μ calculated above this equation gives aggregate income as **£608.5** billion.

If incomes were distributed equally and we assume the increase in utility achieved in speculative markets is negligible and aggregate net borrowing B is zero (both demonstrated above), then aggregate household consumption and therefore total aggregate utility is given by:

$$U = \mu \cdot N \cdot A \cdot e^{\frac{-\mu}{h \cdot \lambda}} \qquad \text{Equation 101}$$

Where   h is the weighted average size of household given in the (Office for National Statistics, 2004) as 2.3 and other terms have their usual meaning

We need to apply the factor **h** to λ because we defined λ for an individual not for a household.

This equation holds despite the gain in utility achieved in speculative markets because the gain is negligible, 27p when **y** $\leq |\lambda|$, the maximum level of useful income for any speculative good **x** is λ_x and $|\lambda_x|$ must always be less than $|\lambda|$. Therefore, the gain in utility in each speculative market must be negligible at this level of income and may be ignored for the purposes of the present analysis.

Even if different consumers were to have different values for λ and the value calculated in chapter 3 was merely its aggregate value or form of average value, equation 101 would still give the maximum likelihood estimate of total utility irrespective of the distribution of λ by the central limit theorem. Equation 101 therefore gives the best obtainable estimate of total utility and subsequent equations give the best estimate of the gain in utility achievable by redistribution. Applying the values of parameters calculated above and those calculated when we defined **V**, then total consumption or utility, if income is distributed equally, can be estimated at **£539.8** billion at year 2000 prices.

Assuming income is distributed as shown in the matrix of co-ordinates on page 211 then aggregate actual income **Y$_A$** is given by:

$$Y_A = \sum_{i=0}^{b-1} n_i \cdot y_i$$

This evaluates to **£608.6** billion, which differs slightly from aggregate income calculated from the mean. The reason is that the matrix on page 211 contains rounding errors, as summing the first row of table 14 (Office for National Statistics, 2004) shows. To eliminate these from our analysis we need to apply a constant **k** to the equation above defined as:

$$k = \frac{608.493}{608.565}$$

Actual aggregate income is then given by:

$$Y_A = k \cdot \sum_{i=0}^{b-1} n_i \cdot y_i \qquad \text{Equation 102}$$

This is quite a subtle way to correct for rounding errors because it spreads the error right across the entire range of incomes. **k** then has the value 0.9999882 and Y_A evaluates to **£608.5** billion, the same value we calculated from the mean.

We can also evaluate total utility **U$_A$** arising from the actual distribution of income as:

$$U_A = k \cdot \sum_{i=0}^{b-1} n_i \cdot y_i \cdot A \cdot e^{\frac{-y_i}{h \cdot \lambda}} \qquad \text{Equation 103}$$

From this equation total aggregate utility arising from the actual distribution of income can be estimated at **£522.9** billion at year 2000 prices. The gain in utility **δU** if incomes were equally distributed can then be calculated from:

$$\delta U = \mu \cdot N \cdot A \cdot e^{\frac{-\mu}{h \cdot \lambda}} - k \cdot \sum_{i=0}^{b-1} n_i \cdot y_i \cdot A \cdot e^{\frac{-y_i}{h \cdot \lambda}} \qquad \text{Equation 104}$$

estimated at **£16.9** billion. This represents over **4.3%** of aggregate consumption or nearly **32** months of the growth of real Nation Income shown in Appendix I table 11. Note however that table 11 is expressed in 1990 prices and table 14 in 1999 – 2000 prices. Even when the figures in table 11 are corrected to year 2000 prices by multiplying by **1.349557** (the proportionate compounded annual average increase in the RPI from 1991 to 2000, calculated from Appendix III table 13) the effect of an equal distribution still represents almost **2** years of real growth shown in table II. This is a significant effect comparable with that of any macro-economic policy measure.

Redistribution of both wealth and income is not Pareto efficient by definition since some consumers will be made worse off so that others can be made better off. Note however that redistribution of income not only improves the wellbeing of the poor but of the very wealthy too. This is because personal incomes above λ result in a loss of total utility. Redistribution of all personal income above λ therefore improves both the wellbeing of the wealthiest consumers as well as the poorest.

The very wealthy may not see it that way however and protest vehemently. Their reaction results from habit because for those that make their own fortunes an increase in income at all levels of income below λ will increase total utility and therefore their wellbeing. The habit of regarding an increase in income as improving their lot will thus have become ingrained. In this the very wealthy are clearly deceiving themselves as their satiating consumption shows.

Many economists would accept redistribution of wealth because much of it is not distributed by the market, for example by inheritance, and therefore its redistribution does not disturb the market's allocative efficiency. Redistribution of income could on the other hand affect allocative efficiency so there are probably practical limits to income redistribution.

It is known that the distribution of income is at least in part structurally determined. The most elegant theoretical analysis of this is to be found in the livelihoods model of (Amartya Sen, 1981) Similarly the correlation coefficients obtained from empirical investigations of some forms of segmented labour market theory suggest there is at least some truth in this (R. McNabb et al, 1981-1996).

We can however apply a simpler argument. The variability of human characteristics and performance are generally normally distributed. Indeed the normal distribution was derived in the first place as a means of predicting how such variables would be distributed. Since labour productivity is an aspect of human performance it is likely to be normally distributed. The first derivative of any normally distributed variable will also be normally distributed so marginal productivity is also likely to be normally distributed.

As even earned, let alone unearned, incomes are not normally distributed we can conclude that the entitlement system of society is distorting allocative efficiency. Basically the powerful abuse their power by rewarding themselves and their supporting hierarchies disproportionately. If marginal productivity were the sole determinant of factor returns, the distribution of earned income would be normal. As a minimum we should therefore redistribute at least until a normal distribution of earned income was achieved. This would actually increase rather than reduce allocative efficiency and provide greater incentives for the great majority of people whilst maintaining society's existing 'pecking order'.

Pecuniary incentives for the wealthy are in any event far less effective than those for lower earners, partly because of their diminishing marginal utility of income but also because rising to the top in any endeavour requires the development of habits of effort and perseverance. The successful are therefore likely to be highly motivated irrespective of the incentives, whereas the less successful need all the incentives we can give them. These arguments about the efficiency of redistribution are the more important reasons for doing it rather than the arguments concerning pursuit of a more equitable distribution for its own sake, important though that may be.

We can calculate and plot the effects of redistributing income normally using the same parameter values calculated above as follows:

$$n_t = \frac{N}{\sigma \cdot \sqrt{2\pi}} \cdot \exp\left[\frac{-1}{2 \cdot \sigma^2}\left(y_t - \mu\right)^2\right]$$

Equation 105

where

n_t is the theoretical frequency or number of households at each level of income assuming a normal distribution

y_t is the corresponding level of income

If we plot such a distribution for the parameter values given above, we obtain a graph of the form shown below.

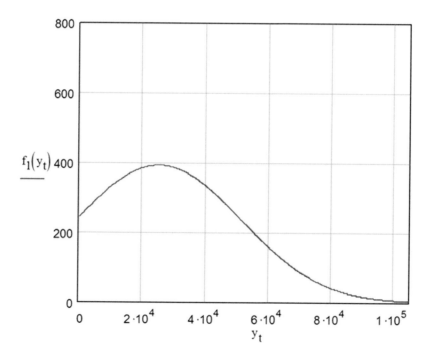

Figure 98. A Normal Distribution of Income with the Same Parameter Values

We can then represent the actual and normal distributions of income together as follows:

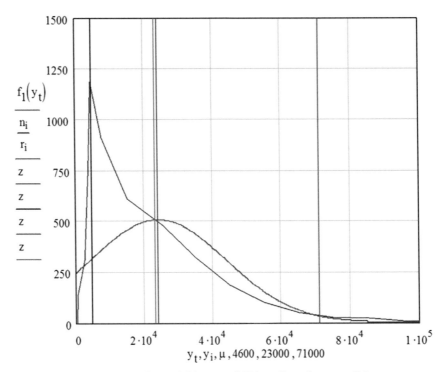

Figure 99. Actual and Normal Distributions of Income

To do this we need to divide the number of households n_i in each income band by r_i the band range or class interval of each band of actual income. This places the discrete actual distribution of income on to an approximately comparable basis with the continuous normal distribution. This is because the class interval of the continuous distribution tends to zero and the number of households within each band would be expected to fall as the class interval becomes smaller. Note the source data shows a slight discontinuity at very low levels of actual income that can be attributed to sampling error. Figure 99 also shows the mean value of income (the vertical brown line) at **£24,019** and values for the points of intersection between the actual distribution of income and the normal distribution at approximately **£23,000** and **£71,000**. Note also average income is **£19,419** greater than modal income at **£4,600**. The most important conclusion from figure 99 is that the number of households outside the range defined above is reduced significantly when incomes are normally distributed. This would greatly reduce the burden on the benefit system. While it is quite simple to estimate the number of households above and below this range (and the change in that number relative to the actual distribution), it is more helpful to consider this cumulatively, so we will defer such an analysis until we have deduced the cumulative effects.

As modal income in figure 99 is increased from **£4,600** per annum to **£24,019** this represents a gain of more than **522%** of modal income. This is proportionally more than all the economic growth shown in Appendix I table 11 or, if we assume growth at the average rate achieved between 1950 and 1992, over 70 years of growth. This represents almost a lifetimes growth of income for the poorest households.

We can calculate aggregate income that would arise under a normal distribution from the equations:

$$Y_t = \int_{-\infty}^{\infty} \frac{N}{\sigma \cdot \sqrt{2 \cdot \pi}} \cdot \exp\left[\frac{-1}{2 \cdot \sigma^2} \cdot (y_t - \mu)^2\right] \cdot y_t \, dy_t$$

or

$$Y_t = \mu \cdot \int_{-\infty}^{\infty} \frac{N}{\sigma \cdot \sqrt{2\pi}} \exp\left[\frac{-1}{2 \cdot \sigma^2} \cdot (y_t - \mu)^2\right] dy_t \qquad \text{Equations 106}$$

Where Y_t is the theoretical level of aggregate income assuming a normal distribution

As we might expect both these equations also evaluate to **£608.5** billion the same value we obtained when calculating aggregate income from the mean or from the data in table 14. From the first of these equations we can calculate the effect of a normal distribution on aggregate utility as follows:

$$U_t = \int_{-\infty}^{\infty} \frac{N}{\sigma \cdot \sqrt{2\pi}} \exp\left[\frac{-1}{2 \cdot \sigma^2} \cdot (y_t - \mu)^2\right] \cdot y_t \cdot A \cdot e^{\frac{-y_t}{h \cdot \lambda}} dy_t \qquad \text{Equation 107}$$

Where **U_t** is aggregate utility arising from a normal distribution of income.

If we assume the same parameter values apply that arose for the actual distribution of income then this equation evaluates to **£522.4** billion. In principle we can then evaluate the net gain or loss of utility relative to the actual distribution of income from the equation:

$$\delta U = \int_{-\infty}^{\infty} \frac{N}{\sigma \cdot \sqrt{2\pi}} \cdot \exp\left[\frac{-1}{2 \cdot (\sigma)^2} (y_t - \mu)^2\right] \cdot y_t \cdot A \cdot e^{\frac{-y_t}{h \cdot \lambda}} dy_t - k \cdot \sum_{i=0}^{b-1} n_i \cdot y_i \cdot A \cdot e^{\frac{-y_i}{h \cdot \lambda}}$$

However because this is a mixed mode calculation involving both discrete array operations and continuous analytical operations MATHCAD cannot evaluate this expression as it stands analytically or numerically. The first term can only be evaluated analytically since ∞ cannot be represented numerically. The second term can only be evaluated numerically because y_i and n_i are discrete array variables (not continuous variables) that can only be evaluated numerically. However since household incomes above **h.λ** result in a net loss of total utility we can restrict the range of the definite integral to ± h.λ. This truncation has little practical effect compared with integration over an infinite interval because for the parametric conditions calculated above:

$$\int_{-\infty}^{\infty} \frac{1}{\sigma \cdot \sqrt{2 \cdot \pi}} \cdot \exp\left[\frac{-1}{2 \cdot \sigma^2} \cdot (y_t - \mu)^2\right] dy_t = 1$$

and

$$\int_{-h \cdot \lambda}^{h \cdot \lambda} \frac{1}{\sigma \cdot \sqrt{2 \cdot \pi}} \cdot \exp\left[\frac{-1}{2 \cdot \sigma^2} \cdot (y_t - \mu)^2\right] dy_t = 1$$

both equal **1** to **19** decimal places. The equation for the net change of utility then takes the form:

$$\delta U = \int_{-h \cdot \lambda}^{h \cdot \lambda} \frac{N}{\sigma \cdot \sqrt{2\pi}} \cdot \exp\left[\frac{-1}{2 \cdot (\sigma)^2} (y_t - \mu)^2\right] \cdot y_t \cdot A \cdot e^{\frac{-y_t}{h \cdot \lambda}} dy_t - k \cdot \sum_{i=0}^{b-1} n_i \cdot y_i \cdot A \cdot e^{\frac{-y_i}{h \cdot \lambda}}$$

Equation 108

This can be evaluated numerically giving a result showing a net loss of utility of **£533.6** million (not billion). This is because, though fewer consumers have very high or very low incomes, more consumers in the middle bands of income would have higher incomes. A normal distribution is therefore more equitable and efficient but not necessarily more egalitarian. However total utility is increased as σ is reduced. The utility neutral break-even point can be calculated by solving the equation given below for σ.

$$\int_{-h \cdot \lambda}^{h \cdot \lambda} \frac{N}{\sigma \cdot \sqrt{2\pi}} \cdot \exp\left[\frac{-1}{2 \cdot (\sigma)^2} (y_t - \mu)^2\right] \cdot A \cdot y_t \cdot e^{\frac{-y_t}{h \cdot \lambda}} dy_t - k \cdot \sum_{i=0}^{b-1} n_i \cdot y_i \cdot A \cdot e^{\frac{-y_i}{h \cdot \lambda}} = 0$$

MATHCAD cannot solve this equation using its own internal iterative processes using either the solve function or the root function but it can be solved iteratively using a simple straddling reducing interval technique as shown below. Note that the equation above is very sensitive to the value of σ so extreme accuracy was sought.

$$\sigma := 19534.453100163, 19534.4531001631 \,..\, 19534.4531001639$$

$$\int_{-h\cdot\lambda}^{h\cdot\lambda} \frac{N}{\sigma\cdot\sqrt{2\pi}}\cdot\exp\left[\frac{-1}{2\cdot(\sigma)^2}(y_t-\mu)^2\right]\cdot A\cdot y_t\cdot e^{\frac{-y_t}{h\cdot\lambda}}\,dy_t - k\cdot\sum_{i=0}^{10} n_i\cdot y_i\cdot A\cdot e^{\frac{-y_i}{h\cdot\lambda}} =$$

0
0
$-3.662109\cdot10^{-4}$
$-4.882813\cdot10^{-4}$
$-6.103516\cdot10^{-4}$
$-8.544922\cdot10^{-4}$
$-9.155273\cdot10^{-4}$
$-1.098633\cdot10^{-3}$
$-1.464844\cdot10^{-3}$

The break-even or utility neutral value of σ is then £19,534.4531001631 and

$$\int_{-h\cdot\lambda}^{h\cdot\lambda} \frac{N}{\sigma\cdot\sqrt{2\cdot\pi}}\cdot\exp\left[\frac{-1}{2\cdot(\sigma)^2}\cdot(y_t-\mu)^2\right]\cdot A\cdot y_t\cdot e^{\frac{-y_t}{h\cdot\lambda}}\,dy_t - k\cdot\sum_{i=0}^{b-1} n_i\cdot y_i\cdot A\cdot e^{\frac{-y_i}{h\cdot\lambda}} = 0$$

This represents a reduction of just under **1%** of the original value of σ. Any further reduction of σ results in a net gain in aggregate utility. Thus, if we desire the distribution of income to be both normal and more egalitarian, we can achieve any desired level of redistribution up to the point of absolute equality of distribution by simply reducing the value of σ. If for example σ were reduced by **23 %** of its utility neutral value, the approximate redistributive effect of the tax and benefit system (excluding pensions) reflected in table 15, below, total utility would increase by approximately **£6.9** billion. This is approximately **41%** of the gain in utility achieved by a completely equal distribution of income and still a very significant 'macro-economic' effect. The result of such a reduction in σ on the distribution of income is shown below.

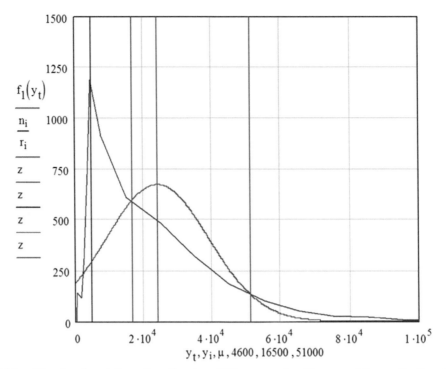

Figure 100. 'Equivalent' Re-distributive Effect as the Tax and Benefit System

The 4 vertical lines show the mean of both the actual and normal distributions in brown, two points of intersection between the actual and normal curves at approximately **£16,500 and £51,000** in magenta and the modal level of actual income at approximately **£4,600**. The normal curve is plotted (with σ calculated on page 213 above reduced by **23%** of its utility neutral value to **£15,079**.

We can deduce the approximate effect of redistribution on income levels by first calculating cumulative frequency and thence deriving the inverse frequency curve as follows. If we assume the truncation of the lower bound of y_t applied above, the cumulative frequency of a normal distribution of income can be calculated and plotted from first principles using the formula:

$$c_t = \int_{-h \cdot \lambda}^{y_t} \frac{N}{\sigma \cdot \sqrt{2 \cdot \pi}} \cdot \exp\left[\frac{-1}{2 \cdot \sigma^2} \cdot (y_t - \mu)^2\right] dy_t \qquad \text{Equation 109}$$

where c_t is the theoretical cumulative frequency assuming a normal distribution

MATHCAD has a function **pnorm** that evaluates cumulative probability for a normal distribution. The theoretical normal cumulative frequency function can therefore also be defined as:

$$c_t = N \cdot \text{pnorm}(y_t, \mu, \sigma)$$

We can demonstrate that these two methods of evaluation yield the same results by plotting the two functions together as shown below. To distinguish the two curves however we need to force them apart because they would otherwise be coincident. This has been achieved by adding £1,000 to y_t and subtracting 1,000 from c_t for the second method of evaluation.

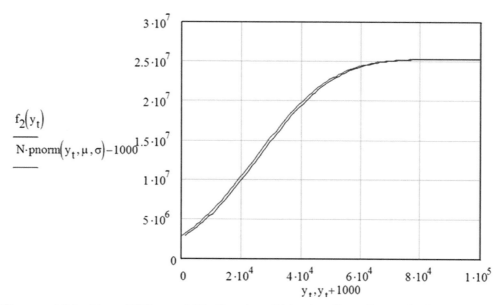

Figure 101. Two Different Methods of Evaluating Cumulative Frequency

We can also define the cumulative frequency c of the actual distribution of incomes using the index variables i as follows:

$$c_i = k \cdot \sum_{i=0}^{b-1} n_i \qquad \text{Equation 110}$$

Thence we can plot the actual and theoretical normal cumulative frequencies, shown in red and blue respectively as shown below assuming the value of μ calculated above and the value of σ for the theoretical distribution reduced by 23% of its utility neutral value. To smooth out the actual cumulative frequency curve at high values of **y** the arbitrary maximum value of **y** of **£105,000** has been increased to **h.λ** its theoretical useful maximum value of approximately **£500,000**. Figure 102 also shows the breakeven level of income and number of households (calculated below) in black.

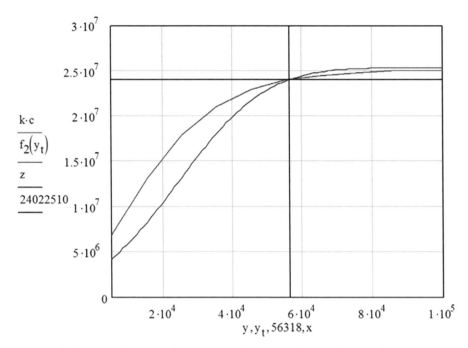

Figure 102. Actual and Normal Cumulative Frequency

The effect on incomes of a normal redistribution with σ reduced by 24% is then given by the 'inverse' of this graph, a graph with the axes exchanged.

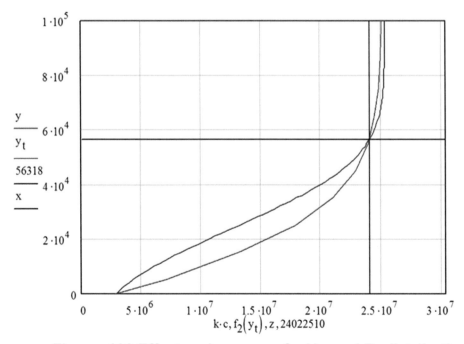

Figure. 103 Effect on Incomes of a Normal Redistribution

In figure 103 the form of the actual inverse cumulative frequency curve in red has been extrapolated for values of y below **£5,000** assuming the actual curve intersects the horizontal axis at the same point as the theoretical normal curve.

Since table 14 provides no information about the distribution of household incomes below **£5,000** this is an arbitrary assumption. However, if we were to make the alternative assumption that the red line intersects the horizontal axis at **c = 0** this introduces an unlikely 'kink' into the curve. Note that this assumption affects only the appearance of the graph and not any of the calculations used in table 10 below or subsequently. We can calculate the intersection of the blue theoretical inverse cumulative frequency curve on the horizontal axis from equation 110 when $y_t = 0$ as **2,863,382**. Under the assumptions that we have made, the actual inverse frequency line is a smooth curve, so the likely point of intersection is around **2.9** million households.

The breakeven level of income of figures 102 and 103 lies between **£55,000** and **£65,000**. We can calculate a reasonable estimate of the breakeven point if we linearly interpolate the actual cumulative frequency curve between these two points by solving the equation:

$$\int_{-h\cdot\lambda}^{y_t} \frac{N}{\sigma\cdot\sqrt{2\cdot\pi}}\cdot\exp\left[\frac{-1}{2\cdot\sigma^2}\cdot(y_t-\mu)^2\right]dy_t - \left[c_6 + (c_7 - c_6)\cdot\frac{y_t - y_6}{y_7 - y_6}\right] = 0$$

MATHCAD cannot solve this equation using its own internal functions but it can be solved iteratively as shown below.

$$y_t := 56351, 56351.1.56352$$

$$\int_{-h\cdot\lambda}^{y_t} \frac{N}{\sigma\cdot\sqrt{2\cdot\pi}}\cdot\exp\left[\frac{-1}{2\cdot\sigma^2}\cdot(y_t-\mu)^2\right]dy_t - \left[c_6 + (c_7 - c_6)\cdot\frac{y_t - y_6}{y_7 - y_6}\right] =$$

-83.864182
-75.472087
-67.080103
-58.688229
-50.296467
-41.904815
-33.513274
-25.121845
-16.730526
-8.339318
0.05178

This gives the break-even value of **y_t** as **£56,352** to the nearest pound and from equation 108 the corresponding value of **c_t** as **24,510,210** households. A more precise conservative estimate will be derived later.

We cannot easily calculate inverse cumulative frequencies from first principles but MATHCAD has a function qnorm that performs this calculation. This function achieves the same effect as exchanging the axes of figure 103, but has the added advantage that it enables us to calculate inverse cumulative frequencies and hence y_t directly. y_t is then given by:

$$y_t = \text{qnorm}(p, \mu, \sigma)$$

where
$$p = \frac{c_t}{N}$$
Equations 111

We can show that this function has an identical effect to the process of exchanging axes used in figure 103 by plotting the two forms of the inverse cumulative frequency curves together with qnorm(**p**, **μ**, **σ**) plotted against **N.p** as shown below. As usual the two curves have been forced apart so they can be distinguished.

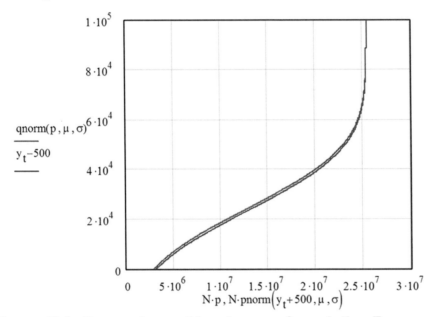

Figure 104. **Comparison of two Inverse Cumulative Frequency Curves**

In principle we now have all the tools needed to calculate the redistributed level of income from the actual level as shown below. For a practical scheme of redistribution however, we would need either a much more finely gradated table of actual incomes (similar to table 14) with income bands of **£1** or preferably **1** penny or we would need a somewhat more finely gradated table, with income bands of say **£100** or **£10**, and a computer program capable of interpolating between points in the table (working in a similar way to the method used to calculate the breakeven point).

To demonstrate the principle we can however simply calculate the redistributed level of income for points in the matrix of coordinates shown on page 211 derived from table 14 of the (Office for National Statistics, 2004).

Actual Income £	Income After Redistribution £	Net Change of Income £	% Change	Estimated Number of Households	Aggregate Net Change £M	Cumulative Aggregate Net Change £M
5,000	11,057	6,057	121.1%	7,022,168	42,533	42,533
15,000	21,174	6,174	41.2%	6,078,280	37,527	80,061
25,000	28,625	3,625	14.5%	4,793,432	17,376	97,437
35,000	34,447	-553	-1.6%	3,172,624	-1,754	95,682
45,000	38,956	-6,044	-13.4%	1,865,779	-11,277	84,406
55,000	42,285	-12,715	-23.1%	1,023,897	-13,019	71,387
65,000	60,170	-4,830	-7.4%	510,940	-2,468	68,919
75,000	63,504	-11,496	-15.3%	276,967	-3,184	65,735
85,000	66,891	-18,109	-21.3%	200,976	-3,639	62,095
95,000	68,802	-26,198	-27.6%	84,990	-2,227	59,869
200,000	95,690	-104,310	-52.1%	300,103	-31,304	28,565

Table 10. Effect of Redistributing Income Normally with σ Reduced by 24%

Note that table 10 suffers from slight rounding errors in the source data. Figures for the estimated number of households are derived from Table 14 except for the estimated number of households in the last row of the table that is calculated from the average proportional decrease in the outstanding number of households per £10,000 of additional income.

The average gain per household (for all those households that gain with incomes below **£35,000**) is **£5,445** and, as column 3 shows, gains are relatively evenly distributed over the first two income bands. In all over **83%** of households totalling over 21 million would benefit from this measure alone. In principle, this table should balance to zero for the economy as a whole. However, we have insufficient information concerning household incomes over **£100,000** at year 2000 prices to prove this. The expressions used to calculate each column of the table, in column order starting from the left (except for the last figure in column 5) are shown below:

$$y_i$$

$$qnorm\left(p_i, \mu, \sigma\right)$$

$$qnorm\left(p_i, \mu, \sigma\right) - y_i$$

$$\frac{\left(qnorm\left(p_i, \mu, \sigma\right) - y_i\right) \cdot \%}{y_i}$$

$$k \cdot n_i$$

$$k \cdot n_i \cdot \left(qnorm\left(p_i, \mu, \sigma\right) - y_i\right)$$

$$\sum_{i=0}^{i} k \cdot n_i \cdot \left(qnorm\left(p_i, \mu, \sigma\right) - y_i\right)$$

for all values of i, where $\quad p_i = \dfrac{1}{N} \cdot \left(c_i - \dfrac{n_i}{2}\right)$

Such a re-distributive scheme 'could' be accomplished through the tax and benefit system. However while these systems have ameliorated the worst privations of poverty and the worst excesses of extreme wealth they could not be regarded as a re-distributive success. The main reason for this is that tax allowances and welfare benefits are based on need. This makes the administrative systems for tax and benefits very complicated. The benefit system is indeed so complicated that even a well-educated, well trained and capable senior civil servant could not be guaranteed to calculate benefits correctly. Furthermore, those most in need of benefits are the least able to cope with this complexity.

For a re-distributive system to work it needs to be simple. This largely precludes all systems based on need. Indeed redistribution should not be approached from the perspective of addressing need but from the perspective of redressing structural distortions in the income distribution (or entitlement) system. When approached in this way the problem becomes much simpler.

A possible scheme is sketched out here. All adults and any children in receipt of any significant amount of money would be required to have a Statutory Bank Account into which all receipts must be paid by law. Any significant receipts paid to an individual that were not paid into his statutory bank account, including all proceeds of crime, would be forfeit to the Crown. Note this requires that all calculations be based on λ not **h.λ**. These accounts should be exempt from all charges. The banks should be required to fund the administration of these simple accounts in consideration of their right to create money when issuing loans. In any event Banks and employers have long desired that all citizens should hold bank accounts and that all benefits, wages salaries and other significant receipts should be paid into a bank. This reduces both the banks and employers risk of theft arising from dealing in and transporting cash.

Some receipts would need to be exempt from redistribution such as small gifts or repayment of small 'un-contracted' informal loans, insurance benefits, proceeds of sale of property, legal damages, criminal compensation, compensation for injuries, normal redundancy payments and some agreed proportion, possibly 60% of (excessive top drawer redundancy or severance payments, National Lottery winnings and inheritance). Exempt payments, except for the first two categories, would need to be made by organisations licensed to make such payments, for example firms of solicitors.

Each bank would then automatically redistribute eligible income using approved computer software and parameters before crediting the recipients' account. The calculations would be similar to those presented here and would be based on a similar, though more finely gradated, matrix of the actual distribution of income. This table could be supplied to the banks by the government or calculated by the banks themselves and should be updated regularly.

The whole process could be very quick (on line real time) and very cheap (per transaction). As the procedure is in principle aggregate income neutral, total receipts into the banking system would meet all entitlements to net income after redistribution, though this would apply to each banking group in total not to individual branches and the reserve bank may need to make and receive payments to balance accounts between commercial banking groups.

Because the calculations are not 'needs based' and because the civil service would not be doing the work this would require no addition government bureaucracy. Indeed many staff now employed to administer tax and benefits systems would no longer be required. The banks may need additional staff but this would be balanced by additional customers, deposits and revenues and judged on a commercial basis. Benefit systems could be much simpler and more generous to the poorest in society because many households that now rely on benefits would no longer need to. Tax laws could be simpler because they would no longer need to be so progressive. As a consequence, income tax could be greatly reduced. For example, the higher rate of tax could be reduced to say **30%**. Despite redistribution this would produce a net benefit to all households with incomes up to **£65,000** at year 2000 prices.

We can estimate the reduced burden on the benefit system from Appendix V Table 15. This applies to the following calculations and all subsequent calculations on page 230. This shows that those households (totalling **5,070,000**), in the lowest quintile of income (with an average gross income of **£5011.2**) receive **78%** of their income in welfare benefits excluding pensions (at an aggregate cost of **£19,817,291,520**) at year 2000 prices.

If such a redistributive policy were then applied, all benefits payments (excluding pensions) paid to the 3rd, 4th and 5th income quintile bands would be entirely unnecessary. The net gain to the exchequer if these benefits were eliminated would be **£29,966,772,420**. Despite these savings the minimum level of household income after redistribution would rise to **£11,057** per annum at year 2000 prices, an increase of over **140.4%** over the modal actual income of **£4,600**. This would require that the level of benefit for all those households solely or substantially dependent on benefit should be raised so that the minimum household income before redistribution was **£5,000**. The total cost to the Treasury of this measure is estimated to be only **£472,524,000**.

Cancelling benefits to higher income bands would of course disturb the original distribution and therefore the levels of redistributed income, but making the distribution of income normal is a much more powerful effect (than redistribution through the tax and benefit system) that would overwhelm the effects of cancelling benefits for higher earners.

Total aggregate personal taxation implied by table 15 amounts to **£117,771,030,000**. The net reduction in benefit payments would allow personal taxation to be reduced by **25.4%** without cost to the exchequer. Possible tax reduction policies could include scrapping the earnings rule applied to all benefits for incomes prior to redistribution up to say **£11,057** (slightly over average actual earnings in the second quintile band at **£10,962**), and raising the minimum tax threshold to the same figure. This would only cost the exchequer approximately **£4,234,464,000**, just over **14.1%** of their net saving from reduced benefits. It would also be possible to rebalance the tax system in favour of direct taxation and away from indirect taxation, a more equitable and egalitarian approach. For example, VAT could be reduced to **10%,** its original rate.

While it may have been true to say in the past that the poor are always with us it certainly need not be so in the future. Furthermore these policies would be so popular that they would be irreversible. We would therefore not only solve the problem of poverty now but permanently!

Introducing such a redistributive scheme is likely to have short-term inflationary effects, which would somewhat defeat its object. These effects would be expected to be short lived because the market would quickly adjust to increased demand for staple goods and goods consumed by most households and reduced demand for conspicuous consumption goods. To reduce such inflationary disturbance to the economy it would be desirable to phase these changes in over say **5** years. Each year the full extent of the intended redistribution of income would be calculated. In the first year **20%** of the redistribution would be applied, in the second year **40%** and so on. Alternatively redistributed income, or a proportion of it, could initially be applied to reducing consumer debt in an irreversible way or funding wider schemes of house purchase or occupational and personal pensions.

Unfortunately if redistribution were phased in, gains to the poorest households would then be quite modest initially. It would therefore probably be desirable to apply a greater proportion of the finally intended redistribution to the lowest bands of income early in the **5** year programme, provided this could be achieved without significant inflationary or other ill effects.

If all the measures discussed above were implemented these policies would be desirable from almost all perspectives. They would be more efficient because income would be proportional to labour productivity, they would be more equitable, they would substantially improve incomes and incentives for approximately **83%** of households, they would be more egalitarian, they would eliminate poverty, they would simplify the tax and benefit system, they would save the exchequer money both directly and in administration costs, they would increase society's total utility significantly and they would comply with the greatest happiness principle.

The only negative effects might be reduced savings and investment, and entrepreneurial flight. The former could be offset by improved fully transferable occupational or personal pensions based on total earning not just basic wages providing substantially better pensions for most people in retirement as well as increased investment from larger pension funds. The government could make flight very uncomfortable for deserting entrepreneurs. For example by making UK citizens liable for redistribution and UK direct taxation (less an offset equal to tax confirmed to have been paid in foreign jurisdictions) no matter where they lived worldwide. To offset any remaining entrepreneurial flight by those that gave up their UK citizenship and the right to visit the UK ever again, since the banks and other financial institutions would have increased deposits at their disposal they could increase lending to business without affecting their fiduciary ratios. The development of long-term lending to, and even investment in, industry by the banks on the German model would in any event be a useful initiative.

The requirement of redistributing earned income to increase society's total utility of income, at least to the point where earned income was normally distributed as described above, would require significant redistribution on a scale not previously attempted. Optimum policies for redistribution of wealth could go even further. There is no obvious reason why all wealth not accumulated from earned income during an individual's lifetime should not be much more fairly distributed, provided that accumulated wealth remained invested. Social welfare would be maximised as a result. Alternatively, individual wealth could be redistributed normally up to the point where the maximum holding was λ/r where **r** is the proportionate average rate of earnings on investment. Practical measures to achieve this might include substantial compulsory redistribution of share capital to employees of companies and all other forms of business, including partnerships and businesses owned by a sole proprietor, extending the right to buy to private tenants, all financed by central not local government, on the same discounted terms as council tenants and making all banks and financial services companies mutual. All shares distributed in this way should be held in trust until normal retirement age, though new shareholders would gain voting rights.

The reason for adopting the last of these redistributive measures is that the general argument for limited liability joint stock companies, that the shareholders take most of the risk and therefore they should gain most of the profits, does not apply to banks and financial services companies. As we saw after the 2008 banking crisis, when these companies fail everyone can lose their money including all depositors and the government, not just the shareholders, because banks create money. If the shareholders are not carrying most of the risk they should not gain most of the profits.

Chapter 13. Conclusions

A new consumption function has been derived using non-linear regression that provides a better fit to available data. From this a simpler and more precise theory of consumer choice and aspects of a new theory of money have been derived. In the process all the axioms of the neoclassical theory of consumer choice have been shown to be unnecessary or incorrect, notably that human wants are satiable not insatiable.

Means of measuring cardinal utility have been proposed and their validity examined and demonstrated. The impracticality of the ordinal utility optimising process and the simplicity of optimisation under the new system has been demonstrated. Of particular importance in this is the idea that the care a buyer needs to take in approximating optimum conditions must be greater where the consequence of error is greater.

Very strong evidence has been assembled attesting to the correctness of the new theory of value. Most particular among this evidence is the fact that the equilibrium conditions of ordinal utility theory are derivable from the results of the new theory. If the new theory were incorrect this would be impossible.

A new theory of value and demand has been developed from these measures of utility and its implications for many different economic problems, investigated. Particularly important results are: the demand function for 'competitive goods', the quadratic solution and other more refined solutions for monopoly price and volume, the price and volume solution for borrowing and the general price and volume solution for oligopoly, which is extendable to all forms of market structure.

The new theory of value and demand also enables an analysis and explanation of speculative markets in which it is demonstrated that such markets are subject to virtual monopoly pricing. Notable results from the speculative market theory include the conditions for speculative markets and conspicuous consumption, a method of predicting the failure of speculative markets and the form of the 'consumption' and demand functions arising in speculative and **Giffen** good markets. The speculative market theory also showed the wide applicability of this analysis to different types of markets including the fashion market.

Important results concerning public welfare have been derived from the new theory. Significant among these results are: that monopoly pricing does not imply a loss of welfare, the greatest happiness of the greatest number is measurable, that there is a maximum level of useful income, and social welfare is maximised if the marginal utility of income of all consumers were equal. Lastly a practical scheme of redistribution and other measures were proposed to achieve a more efficient, equitable and egalitarian normal distribution of income making over **96.5%** of households better off whilst improving economic incentives and efficiency.

Substitution between goods has been analysed using the new theory of value and the endogenous substitution effects of rising income demonstrated. The new theory allows the notions of preference and utility or value to be distinguished. The theory also supports the concept of step function substitution between mutually exclusive goods, which cannot be represented in ordinal utility theory.

An analysis of partial equilibrium derived equations for substitution between all possible combinations of competitive, speculative and **Giffen** goods under the new theory of value. The theory also showed the equivalence of the results of the new theory for cases involving competitive goods to those of ordinal utility theory. Furthermore, it was shown that the equilibrium solution is necessarily always unique in all combinations of markets.

Perhaps the most remarkable result of the new theory of value is the derivation, proof and solution of precise equations for general equilibrium in real markets that demonstrated that the assumptions of the new theory of value are correct and those of ordinal utility theory incorrect. The new theory of value also shows that the unique partial equilibrium solution for each pair of goods holds in a general equilibrium and proposes a much more credible process by which equilibrium is established in practice. Furthermore, this analysis was extended to include: monopoly, oligopoly, speculative and **Giffen** goods, borrowing and money markets and to provide a sufficient approximate solution for goods with discrete demand. The system therefore applies to all goods and services in 'real' markets.

To sum up, the new theory of value has decimated the dominant position of prevailing orthodox views of non-quantifiable and incomparable utility and the whole pre-existing system of ordinal utility theory and general equilibrium analysis by introducing the concept of virtual homogeneous markets and limiting the applicability of the established theory to a much smaller set of cases where these markets apply. I commend the new theory to the profession of economists.

References

Garry Stanley Becker 1981 & 1991, 'A Treatise on the Family', 1991 enlarged edition published by Harvard University Press, Cambridge, Massachusetts and London. Chapter 8 pages 277 to 306 deals with altruism.

Jeremy Bentham, 1789 'An Introduction to the Principles of Morals and Legislation', first published 1789, 1823 edition published by Oxford, Clarendon Press, London, New York and Toronto 1907. 'The Greatest Happiness Principle' is introduced in the first footnote on the first page of chapter 1 as an alternative concept to that of utility.

John Maynard Keynes, 1936, 'The General Theory of Employment, Interest and Money', reprinted 1960, published by Macmillan and Company Ltd, London, 1936. The marginal propensity to consume is introduced in chapter 10 pages 113 to 131.

M. Mackintosh et al, 1996, 'Economics and changing economies' by M. Mackintosh, V. Brown, N. Costello, G. Dawson, G. Thompson and A. Trigg, published by The Open University and International Thomson Business Press.

Alfred Marshal, 1890, 'Principles of Economics', first published 1890, eighth edition 1920, reprinted 1961, published by Macmillan and Company Ltd., London. Marshal considers the satiability of wants, utility, marginal utility and the law of diminishing marginal utility in book III chapter III pages 78 to 85. The idea that gathering data over time affects the form of the demand function is reviewed by Marshal in book III chapter IV section 5 on page 92.

Harold W. Kuhn and Albert W. Tucker, 1951, 'Non-linear programming' published in pages 481 to 492 of 'Proceedings of the Second Berkeley Symposium on Mathematical Statistics and Probability', edited by J. Neyman, published by the University of California Press. The Kuhn-Tucker theorem derives sufficient conditions for non-linear constrained optimisation. According to an article by Tinne Hoff Kjeldsen 'A Contextualized Historical Analysis of the Kuhn-Tucker Theorem in Nonlinear Programming: The Impact of World War II' published in Volume 27, Issue 4 pages 331 to 472 of Historia Mathematica, November 2000, the theorem had been independently proved by William Karush in an unpublished Master's thesis in 1939.

R. McNabb and G. Psacharopoulos, 1981, 'Further evidence on the relevance of dual market labour theory for the UK', Journal of Human Resources, volume 16, number 3, pages 442 to 448.

R. McNabb, 1987, 'Testing for labour market segmentation in Britain', Manchester School, number 3, pages 257 to 273.

R. McNabb and P. Ryan, 1990, 'Segmented labour markets' in 'Current Issues In Labour Economics', edited by D. Sapsford and Z. Tzannatos, London Macmillan and Company Limited.

R. McNabb and K. Whitfield, 1996, 'Labour market segmentation in Britain: evidence from The Third Workplace Industrial Relations Survey' Cardiff Business School Discussion Paper in Economics, 95 – 022.

John Forbes Nash Jr. 1950, 'Non-Cooperative Games', originally presented as a Phd thesis to Princeton University in May 1950. Reprinted in 'The Essential John Nash' edited by Harold W. Kuhn and Sylvia Nasar, published by Princeton University Press, Princeton and Oxford, 2002. Chapter 6 contains a complete statement of the theory in a facsimile of the original Phd thesis and Chapter 7 contains a reprint of a simplified statement of the theory previously published in Annals of Mathematics 54, 1951 pages 286 to 295.

Office for Nation Statistics, 2000 'Family Expenditure Survey 1999-2000', © Crown Copyright 2000. Editor Denis Downs. The source of data for Appendix V is Table 8.3 page 141 of the survey.

Office for National Statistics, 2001. Annual Abstract of Statistics 2001, Editor Keith Tyrrell. Table 5.1 page 26 gives a Population Summary 1964 to 1999 and a projection for 2001. © Crown Copyright 2001. The source data for Appendix II

Office for National Statistics, 2003. Table RP02 gives the Annual Inflation Rate All Items Retail Price Index 1987 to 2003 and Table RP04 gives Annual % Change in Price. © Crown Copyright 2003. The source of data for Appendix III.

Office for National Statistics, 2004 Expenditure and Food Survey 1999-2000, © Crown Copyright 2004. Source of Data for Appendix IV.

Office for National Statistics, 2017. 'Annual Abstract of Statistics 2017' Table 1.2 gives the 'General Index of Retail Prices (All Items RPI), 1946 to 2017, Table 1.4 gives 'Average Earnings of All Full-Time Adult Employees, 1970 to 2017'. Lead Statistician Alan Gibson, Published 11 January 2018. These tables are also available in spreadsheet form from 'The Annual Abstract of Statistics for Benefits and Indices of Prices and Earnings' available on the Internet, accessed on 15 August 2018. © Crown Copyright 2018

OU, 2002, MATHCAD 2001 Professional Worksheet 'Minimizing a function of n variables least squares problems' published in DVD format by The Open University, Milton Keynes, UK as part of the resources for the third level undergraduate mathematics course on Optimization M373.

Vilfredo Federico Damaso Pareto 1909 'Manual of Political Economy', first published 1909, revised edition 1927, translated by Ann S. Schwier, published by Augustus M Kelley, New York 1971. Pareto deals with the general notion of economic equilibrium in chapter III pages 103 to 180 and with economic equilibrium in chapter VI pages 251 to 319. He first asserts that interpersonal comparison of the 'sensations' of one man and another 'is one of the most unsatisfactory in social science' on page 106 section 12.

David Ricardo, 1817, 'The Principles of Political Economy and Taxation' first published in 1817, 1971 edition published by Penguin Books, Harmondsworth, Middlesex, UK. Chapter II pages 91 to 107 introduces the concept of economic rent and chapter VII pages 147 to 167 introduces the concept of comparative advantage.

Amartya Kumar Sen, 1981, 'Poverty and Famines: An Essay on Entitlement and Deprivation', published by Oxford: Clarendon Press.

Adam Smith, 1776 'An Enquiry into the Nature and Causes of the Wealth of Nations', first published 1776, Everyman's Library abridged edition (excluding book V) with introduction by D. D. Raphael, 1991, published by David Campbell Publishers Ltd., Berwick Street, London. Smith considers and rejects utility as the basis of value in book I chapter IV page 24.

Thorstein Bunde Veblen, 1899 'The Theory of the Leisure Class', first published in the UK 1925, 1970 edition published by Unwin Books, London. The concept of conspicuous consumption is introduced in chapter 4, pages 60 to 80 and what Veblen calls 'pecuniary emulation' is considered in chapter 2, pages 33 to 40.

Leon Walras, (Marie-Esprit-Leon Walras) 1874 & 1877, 'Elements of Pure Economics', 1954 edition translated by William Jaffe published by George Allen and Unwin, London. Walras considers partial equilibrium in part II pages 83 to 149 and general equilibrium in part III pages 153 to 207.

Jane Wheelock, 1996, 'People and households as economic agents' published as chapter 3 pages 79 to 111 of 'Economics and changing economies' (M. Mackintosh et al, 1996). Dr Wheelock derives an equation for the aggregate consumption function using linear regression on pages 85 to 89 and provides statistical evidence of UK aggregate consumption between 1950 and 1992 in the form of a table in the appendix to chapter 3 on page 112. The source of Data for Appendix I. The data was originally supplied by the Central Statistical Office, now the Office for National Statistics. © Crown Copyright. The reason for using this source of data rather than more recent data, as elsewhere in this work, is historical. It was this data set that I initially used to overturn the axiom of the insatiability of human wants and hence Vilfredo Pareto's whole system of neoclassical ordinal utility theory. This result was my inspiration for writing this book.

Wikipedia, 2019, 'Lambert W Function' anon. defines both the real and complex forms of Lambert's W Function and its many applications. This article is available on the Internet, last up-dated on 19 June 2019 when accessed on 29 June 2019.

Index

B

C

E

N

T

U

X

X is an arbitrary good · **30**

Y

yield of soft fruit · **30**

yield of the fastest types of computer processor chips is very low · **30**

Z

Z_x is the monopolists profit · **45, 46**

Appendix I Table 11

Personal Income And Expenditure In The UK 1950 To 1992 At 1990 Prices			
Year	Real Income £m	Real Consumption £m	Savings Ratio
1950	125,185	122,649	2.0
1951	123,608	121,042	2.1
1952	126,062	121,088	3.9
1953	132,016	126,362	4.3
1954	136,286	131,587	3.4
1955	142,664	137,136	3.9
1956	146,143	138,105	5.5
1957	148,510	141,004	5.1
1958	150,973	144,614	4.2
1959	158,738	150,913	4.9
1960	169,199	156,735	7.4
1961	176,256	160,199	9.1
1962	178,286	163,925	8.1
1963	185,426	170,874	7.8
1964	193,247	176,044	8.9
1965	196,998	178,493	9.4
1966	201,207	181,550	9.8
1967	204,171	185,985	8.9
1968	207,772	191,209	8.0
1969	209,684	192,366	8.3
1970	217,675	197,873	9.1
1971	220,344	204,139	7.4
1972	238,744	216,752	9.2
1973	254,329	228,615	10.1
1974	252,360	225,317	10.7
1975	253,814	224,580	11.5
1976	253,012	225,666	10.8
1977	247,695	224,892	9.2
1978	265,925	236,909	10.9
1979	281,084	247,212	12.1
1980	285,411	247,185	13.4
1981	283,176	247,402	12.6
1982	281,722	249,852	11.3
1983	289,204	261,200	9.7
1984	299,756	266,486	11.1
1985	309,821	276,742	10.7
1986	323,742	295,622	8.7
1987	334,881	311,234	7.1
1988	354,929	334,591	5.7
1989	372,356	345,406	7.2
1990	380,092	347,527	8.6
1991	378,189	339,993	10.1
1992	388,563	339,610	12.6

Source: Data supplied by the Central Statistical Office now Office for National Statistics, extract.

Appendix II Table 12

Population In Thousands				
Year	United Kingdom	England & Wales	Scotland	Northern Ireland
1964	53,991	47,324	5,209	1,458
1965	54,350	47,671	5,210	1,468
1966	54,643	47,969	5,201	1,476
1967	54,959	48,272	5,198	1,489
1968	55,214	48,511	5,200	1,503
1969	55,461	48,738	5,209	1,514
1970	55,632	48,891	5,214	1,527
1971	55,928	49,152	5,336	1,540
1972	56,097	49,327	5,231	1,539
1973	56,223	49,459	5,234	1,530
1974	56,236	49,468	5,241	1,527
1975	56,226	49,470	5,232	1,524
1976	56,216	49,459	5,233	1,524
1977	56,190	49,440	5,226	1,523
1978	56,178	49,443	5,212	1,523
1979	56,240	49,508	5,204	1,528
1980	56,330	49,603	5,194	1,533
1981	56,357	49,634	5,180	1,543
1982	56,325	49,613	5,167	1,545
1983	56,384	49,681	5,153	1,551
1984	56,513	49,810	5,146	1,557
1985	56,693	49,990	5,137	1,565
1986	56,859	50,162	5,123	1,574
1987	57,015	50,321	5,113	1,582
1988	57,166	50,487	5,093	1,585
1989	57,365	50,678	5,097	1,590
1990	57,567	50,869	5,102	1,596
1991	57,814	51,100	5,107	1,607
1992	58,013	51,277	5,111	1,625
1993	58,198	51,439	5,120	1,638
1994	58,395	51,621	5,132	1,648
1995	58,612	51,820	5,137	1,655
1996	58,807	52,010	5,128	1,669
1997	59,014	52,211	5,123	1,680
1998	59,237	52,428	5,120	1,689
1999	59,501	52,690	5,119	1,692
2000	59,728	52,914	5,114	1,700

Source: Annul Abstract Of Statistics 2001 Table 5.1, extract, published by Office for National Statistics
1964 to 1999 mid-year estimate, 2001 not shown mid-year projection, 2000 linearly interpolated.

Appendix III Table 13

Year	Annual Average Retail Price Index[1]	Annual Average % Change[2]	
1987	101.9	4.2	
1988	106.9	4.9	
1989	115.2	7.8	
1990	126.1	9.5	
1991	133.5	5.9	1.059000
1992	138.5	3.7	1.098183
1993	140.7	1.6	1.115754
1994	144.1	2.4	1.142532
1995	149.1	3.5	1.182521
1996	152.7	2.4	1.210901
1997	157.5	3.1	1.248439
1998	162.9	3.4	1.290886
1999	165.4	1.5	1.310249
2000	170.3	3.0	1.349557
2001	173.3	1.8	
2002	176.2	1.7	
2003	181.3	2.9	

Table title: **Annual Inflation Rate 1987 To 2003**

Source: Office for National Statistics

1. All items monthly Retail Price Index, Table RP02, 2003, extract, January 1987 = 100.

2. Annual % Change, Table RP04, 2003, extract.

3. Figures to the right of the table are calculated cumulative inflation from 1991 to 2000

Appendix IV Table 14

Number of Households by Gross Income Band 1999 - 2000

Annual Income	Under £9,999	£10000 And Under £19,999	£20000 And Under £29,999	£30,000 And Under £39,999	£40,000 and under £49,999	£50,000 and under £59,999	£60,000 and under £69,999	£70,000 and under £79,999	£80,000 and under £89,999	£90,000 and under £99,999	£100,000 or over	All House-holds
Gross households (thousands)	7,023	6,079	4,793	3,173	1,866	1,024	511	277	201	85	304	25,334
Number of households in sample	2,047	1,751	1,336	860	489	267	131	67	50	21	78	7,097
Number of persons in sample	3,387	3,891	3,531	2,552	1,506	822	422	200	153	69	253	16,786
Number of adults in sample	2,547	2,945	2,576	1,833	1,107	609	320	153	118	50	174	12,432
Weighted average number of persons per household	1.6	2.2	2.6	2.9	3.0	3.0	3.2	2.9	3.1	3.4	3.2	2.3

Source: Office for National Statistics, Expenditure and Food Survey 1999-2000, © Crown Copyright 2004

Appendix V Table 15

	United Kingdom Family Expenditure Survey 1999 - 2000								
	Income And Sources Of Income By Gross Income Quintile								
Gross Income Quintile	Gross Number Of Households Thousands	Number Of Households In Sample	Disposable Weekly Household Income £	Gross Weekly Household Income £	Source Of Income Wages And Salaries %	Source Of Income Self Employment %	Source Of Income Investments %	Source Of Income Annuities And Pensions %	Source Of Income Social Security Benefits %	Source Of Income Other %
Lowest Quintile	5,070	1,477	94	96	7	2	4	7	78	3
Second Quintile	5,070	1,483	196	210	27	5	5	14	48	2
Third Quintile	5,070	1,426	315	371	61	7	4	11	15	2
Fourth Quintile	5,070	1,398	468	578	75	6	3	8	6	1
Highest Quintile	5,070	1,313	882	1,145	75	14	5	4	2	1

Source Office for National Statistics, © Crown copyright 2004

Appendix VI Table 16 Excel 20 Good General Equilibrium Simulation

	646	578	761	385	290	822	584	811	586	693	43	720	853	722	491	4	653	309	420	10371
Vc	646	578	761	385	290	822	584	811	586	693	43	720	853	722	491	4	653	309	420	10371
P	50.3400	370.1900	57.0500	113.8300	23.8300	57.1300	36.0800	180.1000	316.5700	49.0100	14.9000	697.3700	48.6500	60.9900	39.2000	0.8500	128.4800	26.4400	154.8700	1.00
D	66.7400	3.5300	70.0100	11.7000	62.5700	76.8300	88.7400	17.3500	4.7300	75.2800	9.3600	1.3100	97.9500	60.3900	64.7900	21.1200	20.4000	59.3900	8.5100	44958
Vs	3360.09	1308.15	3993.78	1331.30	1491.08	4389.43	3201.73	3125.37	1496.23	3689.48	139.46	915.51	4765.36	3683.12	2540.10	17.97	2621.10	1570.50	1318.28	44958
P	50.34	370.19	57.05	113.83	23.83	57.13	36.08	180.10	316.57	49.01	14.90	697.37	48.65	60.99	39.20	0.85	128.48	26.44	154.87	1.00

	646	578	761	385	290	822	584	811	586	693	43	720	853	722	491	4	653	309	420	10371
50.34	1	-0.1360	-0.8825	-0.4423	-2.1126	-0.8812	-1.3954	-0.2795	-0.1590	-1.0272	-3.3787	-0.0722	-1.0348	-0.8254	-1.2842	-59.1878	-0.3918	-1.9037	-0.3251	-50.34
370.19	-7.3533	1	-6.4893	-3.2521	-15.5347	-6.4798	-10.2606	-2.0555	-1.1694	-7.5530	-24.8443	-0.5308	-7.6089	-6.0694	-9.4427	-435.2230	-2.8812	-13.9986	-2.3904	-370.19
57.05	-1.1331	-0.1541	1	-0.5011	-2.3939	-0.9985	-1.5812	-0.3168	-0.1802	-1.1639	-3.8285	-0.0818	-1.1725	-0.9353	-1.4551	-67.0678	-0.4440	-2.1572	-0.3684	-57.05
113.83	-2.2611	-0.3075	-1.9955	1	-4.7769	-1.9925	-3.1551	-0.6321	-0.3596	-2.3225	-7.6396	-0.1632	-2.3397	-1.8663	-2.9036	-133.8300	-0.8860	-4.3046	-0.7350	-113.83
23.83	-0.4733	-0.0644	-0.4177	-0.2093	1	-0.4171	-0.6605	-0.1323	-0.0753	-0.4862	-1.5993	-0.0342	-0.4898	-0.3907	-0.6078	-28.0162	-0.1855	-0.9011	-0.1539	-23.83
57.13	-1.1348	-0.1543	-1.0015	-0.5019	-2.3974	1	-1.5835	-0.3172	-0.1805	-1.1656	-3.8341	-0.0819	-1.1743	-0.9367	-1.4573	-67.1666	-0.4447	-2.1604	-0.3689	-57.13
36.08	-0.7167	-0.0975	-0.6325	-0.3170	-1.5140	-0.6315	1	-0.2003	-0.1140	-0.7361	-2.4213	-0.0517	-0.7416	-0.5915	-0.9203	-42.4171	-0.2808	-1.3643	-0.2330	-36.08
180.10	-3.5773	-0.4865	-3.1570	-1.5821	-7.5575	-3.1524	-4.9917	1	-0.5689	-3.6745	-12.0866	-0.2583	-3.7017	-2.9527	-4.5938	-211.7340	-1.4017	-6.8103	-1.1629	-180.10
316.57	-6.2882	-0.8552	-5.5494	-2.7810	-13.2846	-5.5412	-8.7744	-1.7578	1	-6.4590	-21.2457	-0.4540	-6.5068	-5.1903	-8.0749	-372.1830	-2.4639	-11.9710	-2.0441	-316.57
49.01	-0.9736	-0.1324	-0.8592	-0.4306	-2.0568	-0.8579	-1.3585	-0.2722	-0.1548	1	-3.2893	-0.0703	-1.0074	-0.8036	-1.2502	-57.6224	-0.3815	-1.8534	-0.3165	-49.01
14.90	-0.2960	-0.0403	-0.2612	-0.1309	-0.6253	-0.2608	-0.4130	-0.0827	-0.0471	-0.3040	1	-0.0214	-0.3063	-0.2443	-0.3801	-17.5180	-0.1160	-0.5635	-0.0962	-14.90
697.37	-13.8521	-1.8838	-12.2246	-6.1262	-29.2644	-12.2066	-19.3289	-3.8722	-2.2029	-14.2285	-46.8019	1	-14.3338	-11.4336	-17.7882	-819.8770	-5.4277	-26.3707	-4.5030	-697.37
48.65	-0.9664	-0.1314	-0.8529	-0.4274	-2.0416	-0.8516	-1.3485	-0.2702	-0.1537	-0.9927	-3.2652	-0.0698	1	-0.7977	-1.2410	-57.1990	-0.3787	-1.8398	-0.3142	-48.65
60.99	-1.2115	-0.1648	-1.0692	-0.5358	-2.5595	-1.0676	-1.6905	-0.3387	-0.1927	-1.2444	-4.0934	-0.0875	-1.2537	1	-1.5558	-71.7076	-0.4747	-2.3064	-0.3938	-60.99
39.20	-0.7787	-0.1059	-0.6872	-0.3444	-1.6452	-0.6862	-1.0866	-0.2177	-0.1238	-0.7999	-2.6311	-0.0562	-0.8058	-0.6428	1	-46.0912	-0.3051	-1.4825	-0.2532	-39.20
0.85	-0.0169	-0.0023	-0.0149	-0.0075	-0.0357	-0.0149	-0.0236	-0.0047	-0.0027	-0.0174	-0.0571	-0.0012	-0.0175	-0.0140	-0.0217	1	-0.0066	-0.0321	-0.0055	-0.85
128.48	-2.5521	-0.3471	-2.2523	-1.1287	-5.3917	-2.2490	-3.5612	-0.7134	-0.4059	-2.6215	-8.6228	-0.1842	-2.6409	-2.1065	-3.2773	-151.0550	1	-4.8586	-0.8296	-128.48
26.44	-0.5253	-0.0714	-0.4636	-0.2323	-1.1097	-0.4629	-0.7330	-0.1468	-0.0835	-0.5396	-1.7748	-0.0379	-0.5436	-0.4336	-0.6745	-31.0904	-0.2058	1	-0.1708	-26.44
154.87	-3.0762	-0.4183	-2.7148	-1.3605	-6.4988	-2.7108	-4.2925	-0.8599	-0.4892	-3.1598	-10.3935	-0.2221	-3.1832	-2.5391	-3.9503	-182.0730	-1.2053	-5.8562	1	-154.87
1.00	-0.0199	-0.0027	-0.0175	-0.0088	-0.0420	-0.0175	-0.0277	-0.0056	-0.0032	-0.0204	-0.0671	-0.0014	-0.0206	-0.0164	-0.0255	-1.1757	-0.0078	-0.0378	-0.0065	1

D	67	4	70	12	63	77	89	17	5	75	9	1	98	60	65	21	20	59	9	44958
67	1	-18.8874	-0.9533	-5.7069	-1.0667	-0.8687	-0.7521	-3.8460	-14.1214	-0.8866	-7.1309	-50.8399	-0.6814	-1.1053	-1.0301	-3.1598	-3.2717	-1.1239	-7.8408	-0.0015
4	-0.0530	1	-0.0505	-0.3022	-0.0565	-0.0460	-0.0398	-0.2036	-0.7477	-0.0469	-0.3775	-2.6917	-0.0361	-0.0585	-0.0545	-0.1673	-0.1732	-0.0595	-0.4151	-0.0001
70	-1.0489	-19.8118	1	-5.9862	-1.1189	-0.9112	-0.7889	-4.0342	-14.8125	-0.9300	-7.4798	-53.3281	-0.7148	-1.1594	-1.0805	-3.3145	-3.4318	-1.1789	-8.2245	-0.0016
12	-0.1752	-3.3096	-0.1671	1	-0.1869	-0.1522	-0.1318	-0.6739	-2.4745	-0.1554	-1.2495	-8.9086	-0.1194	-0.1937	-0.1805	-0.5537	-0.5733	-0.1969	-1.3739	-0.0003
63	-0.9375	-17.7070	-0.8938	-5.3502	1	-0.8144	-0.7051	-3.6056	-13.2389	-0.8312	-6.6852	-47.6625	-0.6388	-1.0362	-0.9657	-2.9624	-3.0672	-1.0536	-7.3507	-0.0014
77	-1.1512	-21.7425	-1.0975	-6.5695	-1.2279	1	-0.8658	-4.4274	-16.2560	-1.0207	-8.2088	-58.5249	-0.7844	-1.2723	-1.1858	-3.6375	-3.7662	-1.2937	-9.0260	-0.0017
89	-1.3296	-25.1130	-1.2676	-7.5879	-1.4183	-1.1550	1	-5.1137	-18.7760	-1.1789	-9.4813	-67.5973	-0.9060	-1.4696	-1.3697	-4.2014	-4.3500	-1.4943	-10.4252	-0.0020
17	-0.2600	-4.9109	-0.2479	-1.4839	-0.2773	-0.2259	-0.1956	1	-3.6717	-0.2305	-1.8541	-13.2189	-0.1772	-0.2874	-0.2678	-0.8216	-0.8507	-0.2922	-2.0387	-0.0004
5	-0.0708	-1.3375	-0.0675	-0.4041	-0.0755	-0.0615	-0.0533	-0.2724	1	-0.0628	-0.5050	-3.6002	-0.0483	-0.0783	-0.0730	-0.2238	-0.2317	-0.0796	-0.5552	-0.0001
75	-1.1279	-21.3024	-1.0752	-6.4365	-1.2031	-0.9798	-0.8483	-4.3377	-15.9270	1	-8.0426	-57.3402	-0.7685	-1.2466	-1.1618	-3.5639	-3.6900	-1.2675	-8.8433	-0.0017
9	-0.1402	-2.6487	-0.1337	-0.8003	-0.1496	-0.1218	-0.1055	-0.5393	-1.9803	-0.1243	1	-7.1296	-0.0956	-0.1550	-0.1445	-0.4431	-0.4588	-0.1576	-1.0996	-0.0002
1	-0.0197	-0.3715	-0.0188	-0.1123	-0.0210	-0.0171	-0.0148	-0.0757	-0.2778	-0.0174	-0.1403	1	-0.0134	-0.0217	-0.0203	-0.0622	-0.0644	-0.0221	-0.1542	0.0000
98	-1.4675	-27.7180	-1.3991	-8.3750	-1.5654	-1.2748	-1.1037	-5.6441	-20.7237	-1.3012	-10.4648	-74.6094	1	-1.6220	-1.5117	-4.6372	-4.8013	-1.6493	-11.5066	-0.0022
60	-0.9048	-17.0885	-0.8625	-5.1633	-0.9651	-0.7860	-0.6805	-3.4797	-12.7764	-0.8022	-6.4517	-45.9977	-0.6165	1	-0.9320	-2.8589	-2.9601	-1.0168	-7.0940	-0.0013
65	-0.9708	-18.3353	-0.9255	-5.5400	-1.0355	-0.8433	-0.7301	-3.7336	-13.7086	-0.8607	-6.9224	-49.3537	-0.6615	-1.0730	1	-3.0675	-3.1760	-1.0910	-7.6115	-0.0014
21	-0.3165	-5.9774	-0.3017	-1.8061	-0.3376	-0.2749	-0.2380	-1.2172	-4.4690	-0.2806	-2.2567	-16.0894	-0.2157	-0.3498	-0.3260	1	-1.0354	-0.3557	-2.4814	-0.0005
20	-0.3057	-5.7730	-0.2914	-1.7443	-0.3260	-0.2655	-0.2299	-1.1755	-4.3163	-0.2710	-2.1796	-15.5395	-0.2083	-0.3378	-0.3149	-0.9658	1	-0.3435	-2.3966	-0.0002
59	-0.8898	-16.8060	-0.8483	-5.0780	-0.9491	-0.7730	-0.6692	-3.4222	-12.5652	-0.7889	-6.3450	-45.2373	-0.6063	-0.9835	-0.9166	-2.8116	-2.9111	1	-6.9767	-0.0013
9	-0.1275	-2.4089	-0.1216	-0.7278	-0.1360	-0.1108	-0.0959	-0.4905	-1.8010	-0.1131	-0.9095	-6.4841	-0.0869	-0.1410	-0.1314	-0.4030	-0.4173	-0.1433	1	-0.0002
44958	-673.60	-12722.50	-642.17	-3844.13	-718.50	-585.15	-506.61	-2590.65	-9512.17	-597.24	-4803.33	-3425.70	-459.00	-744.51	-693.88	-2128.46	-2203.79	-757.02	-5281.52	1

Description of the Model for a 20 Good General Equilibrium Simulation

The price P of any of the first 19 goods in the 20 good matrix is calculated from any of the following equations randomly selected:

$P = V/D$ where $D = V/C_D$ and $C_D = C_F/D + C_V + C_E D$ for goods in a Competitive market

$P = C_D \log_e D + V/D$ where $D \geq V/C_D$ and $D \leq 2V/C_D$ for goods in a Speculative market

$P = V - C_D \log_e D - V/D$ where $D \geq 1$ and $D \leq V/C_D$ for goods in a **Giffen** good market

$P = V/D$ where D is calculated from $\dfrac{d^2(f_7(D))}{dD^2} = 0$ for maximum profitability

or $D = \sqrt{C_F/C_E}$ for minimum cost for goods in Monopoly or Oligopoly markets

The price of the 20th good – money is £1

The value of the parameters V, C_D, C_F, C_V and C_E must also be randomly selected from within a specified range for each parameter

The value of the variable D must be randomly selected from within a specified range for goods in a Competitive market

The value of the variable D must be randomly selected from within the ranges defined above for Speculative goods, **Giffen** goods, Monopoly goods or Oligopoly goods.

Note each time the simulation is run new prices and levels of demand are selected for each row and column.

Though MATHCAD supports 2 methods of solving systems of equations expressed in matrix form, Isolve and Find, that are more accurate and efficient than matrix inversion, particularly for very large systems of equations, Matrix inversion is however the definitive way to test that a matrix is linearly independent, non-singular and therefore determinant. We will therefore solve our 20-good simulation using matrix inversion.

We begin the solution process by extracting two 10 good sub-matrices from the upper P Matrix of the 20 good simulation matrix, the first 10 goods are extracted from the top left hand corner of the P matrix and the second 10 goods are extracted from the bottom right hand corner of the P matrix in Appendix VI Table 16 Page 263. The two 10 good matrices are shown on the next page and their inverted matrices thereafter.

The two 10 good systems of equations are then solved by the procedure developed in chapter 11 from equations 91, 92 and 94 on pages 192 to 196.

Let D_1 be the sub-matrix for the first 10 goods:

$$D_1 := \begin{pmatrix}
1 & -0.13599 & -0.88251 & -0.44226 & -2.11263 & -0.88121 & -1.39538 & -0.27954 & -0.15903 & -1.02717 \\
-7.35325 & 1 & -6.4893 & -3.25205 & -15.5347 & -6.47976 & -10.2606 & -2.05552 & -1.16938 & -7.55302 \\
-1.13313 & -0.1541 & 1 & -0.50114 & -2.39389 & -0.99853 & -1.58115 & -0.31675 & -0.1802 & -1.16392 \\
-2.26111 & -0.3075 & -1.99545 & 1 & -4.77689 & -1.99252 & -3.15511 & -0.63207 & -0.35958 & -2.32254 \\
-0.47334 & -0.06437 & -0.41773 & -0.20934 & 1 & -0.41712 & -0.66049 & -0.13232 & -0.07528 & -0.4862 \\
-1.1348 & -0.15433 & -1.00147 & -0.50188 & -2.39742 & 1 & -1.58348 & -0.31722 & -0.18047 & -1.16563 \\
-0.71665 & -0.09746 & -0.63245 & -0.31695 & -1.51402 & -0.63152 & 1 & -0.20033 & -0.11397 & -0.73612 \\
-3.57732 & -0.4865 & -3.15701 & -1.58211 & -7.55754 & -3.15237 & -4.99171 & 1 & -0.5689 & -3.67451 \\
-6.28817 & -0.85516 & -5.54936 & -2.78101 & -13.2846 & -5.5412 & -8.77437 & -1.75779 & 1 & -6.45901 \\
-0.97355 & -0.1324 & -0.85917 & -0.43056 & -2.05675 & -0.8579 & -1.35847 & -0.27215 & -0.15482 & 1
\end{pmatrix}$$

and D_2 be the sub-matrix for the second 10 goods

$$D_2 := \begin{pmatrix}
1 & -0.02137 & -0.30626 & -0.2443 & -0.38007 & -17.518 & -0.11597 & -0.56345 & -0.09621 & -14.9004 \\
-46.8019 & 1 & -14.3338 & -11.4336 & -17.7882 & -819.877 & -5.42768 & -26.3707 & -4.50301 & -697.367 \\
-3.26515 & -0.06977 & 1 & -0.79767 & -1.241 & -57.199 & -0.37866 & -1.83976 & -0.31415 & -48.652 \\
-4.09336 & -0.08746 & -1.25365 & 1 & -1.55578 & -71.7076 & -0.47471 & -2.30642 & -0.39384 & -60.9927 \\
-2.63107 & -0.05622 & -0.8058 & -0.64277 & 1 & -46.0912 & -0.30513 & -1.48249 & -0.25315 & -39.204 \\
-0.05708 & -0.00122 & -0.01748 & -0.01395 & -0.0217 & 1 & -0.00662 & -0.03216 & -0.00549 & -0.85057 \\
-8.62281 & -0.18424 & -2.64086 & -2.10654 & -3.2773 & -151.055 & 1 & -4.85856 & -0.82964 & -128.483 \\
-1.77477 & -0.03792 & -0.54355 & -0.43357 & -0.67454 & -31.0904 & -0.20582 & 1 & -0.17076 & -26.4448 \\
-10.3935 & -0.22207 & -3.18315 & -2.5391 & -3.95028 & -182.073 & -1.20534 & -5.85624 & 1 & -154.867 \\
-0.06711 & -0.00143 & -0.02055 & -0.0164 & -0.02551 & -1.17568 & -0.00778 & -0.03781 & -0.00646 & 1
\end{pmatrix}$$

Then:

$D_1{}^{-1} =$

$$\begin{pmatrix}
0.437501332 & -0.008500591 & -0.055155638 & -0.027641055 & -0.132037591 & -0.055075279 & -0.087209526 & -0.017470767 & -0.009939299 & -0.064196386 \\
-0.459572914 & 0.437500291 & -0.405581403 & -0.203253199 & -0.970914233 & -0.404985692 & -0.641274336 & -0.128470076 & -0.073088718 & -0.472061233 \\
-0.070820096 & -0.009630995 & 0.437500988 & -0.03132086 & -0.149616433 & -0.062407186 & -0.098820332 & -0.019798214 & -0.01126309 & -0.072743143 \\
-0.141318336 & -0.019218126 & -0.124715252 & 0.437500035 & -0.298555389 & -0.124531188 & -0.197192501 & -0.03950405 & -0.022475244 & -0.145158206 \\
-0.029584989 & -0.0040238 & -0.02610834 & -0.013084281 & 0.437499307 & -0.026068965 & -0.041282158 & -0.008269501 & -0.00470378 & -0.030388799 \\
-0.070926124 & -0.009644692 & -0.062593531 & -0.031367397 & -0.149839957 & 0.437498974 & -0.098968353 & -0.019827061 & -0.011278847 & -0.072853395 \\
-0.044790615 & -0.006091385 & -0.039528319 & -0.019808287 & -0.094625807 & -0.039470318 & 0.437500206 & -0.012521275 & -0.007122861 & -0.046007655 \\
-0.223581016 & -0.030404716 & -0.19731493 & -0.098881235 & -0.472351032 & -0.197024827 & -0.311983653 & 0.43749955 & -0.035556876 & -0.229655831 \\
-0.39300704 & -0.053445871 & -0.346835481 & -0.173812786 & -0.830279083 & -0.346326509 & -0.54839917 & -0.109861754 & 0.437497316 & -0.403685103 \\
-0.060846634 & -0.00827416 & -0.053697106 & -0.026910972 & -0.128547662 & -0.053619967 & -0.084904857 & -0.017008002 & -0.009677487 & 0.437500185
\end{pmatrix}$$

Is the inverted sub-matrix for the first 10 goods and

$D_2{}^{-1} =$

$$\begin{pmatrix}
0.437490522 & -0.001333314 & -0.019142935 & -0.015275894 & -0.023763414 & -1.095107667 & -0.007248372 & -0.035219377 & -0.006015655 & -0.931460195 \\
-2.925455612 & 0.437561516 & -0.895830525 & -0.714940806 & -1.112081877 & -51.25092083 & -0.339209296 & -1.648231247 & -0.281484004 & -43.59252424 \\
-0.204099711 & -0.004354966 & 0.437500188 & -0.04987886 & -0.077586318 & -3.575610726 & -0.023666523 & -0.114991987 & -0.019639236 & -3.041321166 \\
-0.255861592 & -0.005461248 & -0.078350796 & 0.437471289 & -0.097263402 & -4.482426218 & -0.029668021 & -0.144154634 & -0.024618608 & -3.812620369 \\
-0.164466721 & -0.003509531 & -0.050364249 & -0.040192023 & 0.437479136 & -2.881274403 & -0.019069761 & -0.092661192 & -0.015823911 & -2.450734478 \\
-0.00356907 & -7.60774E\text{-}05 & -0.001093324 & -0.00087086 & -0.001355406 & 0.437491371 & -0.000413748 & -0.002011333 & -0.000343889 & -0.053169286 \\
-0.538987695 & -0.011503853 & -0.16505032 & -0.131719445 & -0.204892778 & -9.442407359 & 0.437504125 & -0.303669491 & -0.051860095 & -8.031590896 \\
-0.110933662 & -0.0023679 & -0.033970148 & -0.02711124 & -0.042171414 & -1.943459361 & -0.012863484 & 0.437498708 & -0.010673454 & -1.653029699 \\
-0.649651571 & -0.0138667 & -0.198940463 & -0.15876807 & -0.246963961 & -11.38137198 & -0.075329786 & -0.366024059 & 0.437490548 & -9.680602093 \\
-0.004189897 & -9.04047E\text{-}05 & -0.001283894 & -0.001022659 & -0.001591943 & -0.073391328 & -0.000486532 & -0.002361473 & -0.000402382 & 0.43757445
\end{pmatrix}$$

is the inverted sub-matrix for the second 10 goods

Note that the terms on the leading diagonals of both inverted matrices are approximately equal to 0.4375 to 4 significant figures, which equates to $(n - 3)/2(n - 2)$ where $n = 10$. Every other term in both inverted matrices equates to $- P_r/(2n - 2)P_x$ where P_r is the 'row' price shown to the left of each row of the model and P_x is the 'column' price shown at the head of each column of the model. Each figure in the matrices D_1 and D_2, apart from those on the leading diagonal, is therefore divided by $(2n - 2)$, which evaluates to 16 for $n = 10$, in the inverted matrices $D_1{}^{-1}$ and $D_2{}^{-1}$.

From Equation 96 the vector $\underline{V}$ evaluates to:

$$-8 \cdot \begin{pmatrix} 50.34 \\ 370.19 \\ 57.05 \\ 113.83 \\ 23.83 \\ 57.13 \\ 36.08 \\ 180.1 \\ 316.57 \\ 49.01 \end{pmatrix} \rightarrow \begin{pmatrix} -402.72 \\ -2961.52 \\ -456.4 \\ -910.64 \\ -190.64 \\ -457.04 \\ -288.64 \\ -1440.8 \\ -2532.56 \\ -392.08 \end{pmatrix}$$

for the first 10 goods and

$$-8 \begin{pmatrix} 14.90 \\ 697.37 \\ 48.65 \\ 60.99 \\ 39.29 \\ 0.85 \\ 128.48 \\ 26.44 \\ 154.87 \\ 1 \end{pmatrix} \rightarrow \begin{pmatrix} -119.2 \\ -5578.96 \\ -389.2 \\ -487.92 \\ -314.32 \\ -6.8 \\ -1027.84 \\ -211.52 \\ -1238.96 \\ -8 \end{pmatrix}$$

for the second 10 goods

The vector $\underline{P}$ then evaluates to:

$$
\begin{pmatrix}
0.437501332 & -0.008500591 & -0.055155638 & -0.027641055 & -0.132037591 & -0.055075279 & -0.087209526 & -0.017470767 & -0.009939299 & -0.064196386 \\
-0.459572914 & 0.437500291 & -0.405581403 & -0.203253199 & -0.970914233 & -0.404985692 & -0.641274336 & -0.128470076 & -0.073088718 & -0.472061233 \\
-0.070820096 & -0.009630995 & 0.437500988 & -0.03132086 & -0.149616433 & -0.062407186 & -0.098820332 & -0.019798214 & -0.01126309 & -0.072743143 \\
-0.141318336 & -0.019218126 & -0.124715252 & 0.437500035 & -0.298555389 & -0.124531188 & -0.197192501 & -0.03950405 & -0.022475244 & -0.145158206 \\
-0.029584989 & -0.0040238 & -0.02610834 & -0.013084281 & 0.437499307 & -0.026068965 & -0.041282158 & -0.008269501 & -0.00470378 & -0.030388799 \\
-0.070926124 & -0.009644692 & -0.062593531 & -0.031367397 & -0.149839957 & 0.437498974 & -0.098968353 & -0.019827061 & -0.011278847 & -0.072853395 \\
-0.044790615 & -0.006091385 & -0.039528319 & -0.019808287 & -0.094625807 & -0.039470318 & 0.437500206 & -0.012521275 & -0.007122861 & -0.046007655 \\
-0.223581016 & -0.030404716 & -0.19731493 & -0.098881235 & -0.472351032 & -0.197024827 & -0.311983653 & 0.43749955 & -0.035556876 & -0.229655831 \\
-0.39300704 & -0.053445871 & -0.346835481 & -0.173812786 & -0.830279083 & -0.346326509 & -0.54839917 & -0.109861754 & 0.437497316 & -0.403685103 \\
-0.060846634 & -0.00827416 & -0.053697106 & -0.026910972 & -0.128547662 & -0.053619967 & -0.084904857 & -0.017008002 & -0.009677487 & 0.437500185
\end{pmatrix}
\begin{pmatrix}
-402.72 \\ -2961.52 \\ -456.4 \\ -910.64 \\ -190.64 \\ -457.04 \\ -288.64 \\ -1440.8 \\ -2532.56 \\ -392.08
\end{pmatrix}
\rightarrow
\begin{pmatrix}
50.35749798264 \\ 370.18537937768 \\ 57.0294375548 \\ 113.84253953008 \\ 23.82946336368 \\ 57.13113995576 \\ 36.0741040152 \\ 180.07668091976 \\ 316.56627622792 \\ 49.02042063752
\end{pmatrix}
$$

for the first 10 goods and

$$
\begin{pmatrix}
0.437490522 & -0.001333314 & -0.019142935 & -0.015275894 & -0.023763414 & -1.095107667 & -0.007248372 & -0.035219377 & -0.006015655 & -0.931460195 \\
-2.925455612 & 0.437561516 & -0.895830525 & -0.714940806 & -1.112081877 & -51.25092083 & -0.339209296 & -1.648231247 & -0.281484004 & -43.59252424 \\
-0.204099711 & -0.004354966 & 0.437500188 & -0.04987886 & -0.077586318 & -3.575610726 & -0.023666523 & -0.114991987 & -0.019639236 & -3.041321166 \\
-0.255861592 & -0.005461248 & -0.078350796 & 0.437471289 & -0.097263402 & -4.482426218 & -0.029668021 & -0.144154634 & -0.024618608 & -3.812620369 \\
-0.164466721 & -0.003509531 & -0.050364249 & -0.040192023 & 0.437479136 & -2.881274403 & -0.019069761 & -0.092661192 & -0.015823911 & -2.450734478 \\
-0.00356907 & -7.60774E\text{-}05 & -0.001093324 & -0.00087086 & -0.001355406 & 0.437491371 & -0.000413748 & -0.002011333 & -0.000343889 & -0.053169286 \\
-0.538987695 & -0.011503853 & -0.16505032 & -0.131719445 & -0.204892778 & -9.442407359 & 0.437504125 & -0.303669491 & -0.051860095 & -8.031590896 \\
-0.110933662 & -0.0023679 & -0.033970148 & -0.02711124 & -0.042171414 & -1.943459361 & -0.012863484 & 0.437498708 & -0.010673454 & -1.653029699 \\
-0.649651571 & -0.0138667 & -0.198940463 & -0.15876807 & -0.246963961 & -11.38137198 & -0.075329786 & -0.366024059 & 0.437490548 & -9.680602093 \\
-0.004189897 & -9.04047E\text{-}05 & -0.001283894 & -0.001022659 & -0.001591943 & -0.073391328 & -0.000486532 & -0.002361473 & -0.000402382 & 0.43757445
\end{pmatrix}
\begin{pmatrix}
-119.2 \\ -5578.96 \\ -389.2 \\ -487.92 \\ -314.32 \\ -6.8 \\ -1027.84 \\ -211.52 \\ -1238.96 \\ -8
\end{pmatrix}
\rightarrow
\begin{pmatrix}
14.91413495592 \\ 697.89748094512 \\ 48.69907253184 \\ 61.05025558984 \\ 38.89187477144 \\ 0.853510422544 \\ 128.59651451296 \\ 26.48394905832 \\ 154.97292209056 \\ 0.999423228312
\end{pmatrix}
$$

for the second 10 goods

The table below shows the actual price of each good chosen in the simulation and the calculated solutions for the price of each good to 1p derived from the equations listed above and the resulting % error.

Price £ chosen by the Simulation	Calculated Price £	% Error
50.34	50.36	-0.040%
370.19	370.19	0.000%
57.05	57.03	0.035%
113.83	113.84	-0.009%
23.83	23.83	0.000%
57.13	57.13	0.000%
36.08	36.07	0.028%
180.1	180.08	0.011%
316.57	316.57	0.000%
49.01	49.02	-0.020%
14.9	14.91	-0.067%
697.37	697.90	-0.076%
48.65	48.70	-0.103%
60.99	61.05	-0.098%
39.29	38.89	1.018%
0.85	0.85	0.000%
128.48	128.60	-0.093%
26.44	26.48	-0.151%
154.87	154.97	-0.065%
1	1	0.000%
Average Modulus of % Error		**0.001987%**

Table 17. Summary of General Equilibrium Simulation Errors